Nanoscience Applications in Diabetes Treatment

Edited by

Ali Rastegari

*Department of Pharmaceutics and Pharmaceutical
Nanotechnology, School of Pharmacy
Iran University of Medical Sciences
Tehran, Iran*

Nanoscience Applications in Diabetes Treatment

Editor: Ali Rastegari

ISBN (Online): 978-981-5196-53-5

ISBN (Print): 978-981-5196-54-2

ISBN (Paperback): 978-981-5196-55-9

First published in 2023.

need for a court order if at any point you breach any terms of this License Agreement. In no event will any delay or failure by Bentham Science Publishers in enforcing your compliance with this License Agreement constitute a waiver of any of its rights.

3. You acknowledge that you have read this License Agreement, and agree to be bound by its terms and conditions. To the extent that any other terms and conditions presented on any website of Bentham Science Publishers conflict with, or are inconsistent with, the terms and conditions set out in this License Agreement, you acknowledge that the terms and conditions set out in this License Agreement shall prevail.

Bentham Science Publishers Pte. Ltd.
80 Robinson Road #02-00
Singapore 068898
Singapore
Email: subscriptions@benthamscience.net

BENTHAM SCIENCE

CONTENTS

PREFACE

The book is structured in a manner that sequentially covers various aspects related to diabetes and the application of nanotechnology in its treatment. Chapters 1 and 2 extensively delve into the pathophysiology of diabetes, encompassing different types of the disease, and provide an overview of the diverse medical therapy approaches available for each type. Chapter 3 focuses on the utilization of nanomedicine for insulin delivery in diabetes treatment. It thoroughly explores the various nano-based vehicles that hold the potential for delivering insulin effectively. In Chapter 4, the book extensively discusses the potential of nanoscience in drug delivery for diabetes. This chapter presents a comprehensive review of different studies that have investigated the use of nanoparticles as carriers for drug delivery in diabetes treatment. The final chapter concentrates on nanotechnology approaches for nucleotide delivery and gene therapy in diabetes. It not only highlights the advancements in this field but also addresses the associated challenges and potential future developments. Overall, the book aims to provide a comprehensive understanding of diabetes, current medical therapies, and how nanotechnology can be harnessed to enhance treatment options, including insulin delivery, drug delivery, and gene therapy.

Ali Rastegari
Department of Pharmaceutics and Pharmaceutical Nanotechnology
School of Pharmacy, Iran University of Medical Sciences
Tehran, Iran

List of Abbreviations

AGIs	Alpha-glucosidase inhibitors
BMI	Body mass index
CDC	Centers for disease control and prevention
CPPs	Cell-penetrating peptides
CNTs	Carbon nanotubes
DM	Diabetes mellitus
DDs	Drug delivery systems
DPPi	Dipetidyl peptide inhibitor
DNA	Deoxyribonucleic acid
FDA	Food and drug administration
G-CSF	Granulocyte colony-stimulating factor
GAD	Glutamic acid decarboxylase
GI	Gastrointestinal
GLP-1	Glucagon-like peptide-1
GIP	Gastric inhibitory polypeptide
GCK	Glucokinase
GLB	Glibenclamide
HLA	Human leukocyte antigen
IL	Interleukin
ICAs	Islet cell antibodies
IAAs	Insulin autoantibodies
LPL	Lipoprotein lipase
LADA	Latent autoimmune diabetes of adults
LNCs	Lipid nano-capsules
MNPs	Metallic nanoparticles
MODY	Maturity-onset Diabetes of the Young
MCP1	Monocyte chemoattractant protein 1
MEs	Micro-emulsions
NPs	Nanoparticles
NEs	Nano-emulsions
NLCs	Nanostructured Lipid Carriers
PKC	Protein kinase C

PLs   Pro-liposomes

PEG   Poly ethylene glycol

PLGA   Poly (lactic-co-glycolic) acid

RAA   Rapid-acting analogs

RNA   Ribonucleic acid

SiRNA   Short interfering RNAs

SGDC   Sodium-glycodeoxycholate

SLNs   Solid lipid NPs

SUR   Sulfonylurea receptor

SGLT2   Sodium-glucose cotransporter 2

TCA   Tricarboxylic acid

TZD   Thiazolidinediones

TNF   Tumor necrosis factor

T1DM   Type 1 diabetes mellitus

T2DM   Type 2 diabetes mellitus

UCP1   Uncoupling protein 1

List of Contributors

Ali Rastegari	Department of Pharmaceutics and Pharmaceutical Nanotechnology, School of Pharmacy, Iran University of Medical Sciences, Tehran, Iran
Farid Abedin Dorkoosh	Faculty of Pharmacy, Tehran University of Medical Sciences, Tehran, Iran
Hitesh Kumar	Department of Pharmaceutics, JSS College of Pharmacy, JSS Academy of Higher Education & Research, Mysuru, Karnataka, India
K. Trideva Sastri	Department of Pharmaceutics, JSS College of Pharmacy, JSS Academy of Higher Education & Research, Mysuru, Karnataka, India
M. Sharadha	Department of Pharmaceutics, JSS College of Pharmacy, JSS Academy of Higher Education & Research, Mysuru, Karnataka, India
Mohammad Vaziri	Faculty of Pharmacy, Tehran University of Medical Sciences, Tehran, Iran
Morteza Rafiee-Tehrani	Faculty of Pharmacy, Tehran University of Medical Sciences, Tehran, Iran
N. Vishal Gupta	Department of Pharmaceutics, JSS College of Pharmacy, JSS Academy of Higher Education & Research, Mysuru, Karnataka, India
Neda Hatami	Endocrine Research Center, Institute of Endocrinology and Metabolism, Iran University of Medical Sciences, Tehran, Iran
Ramin Malboosbaf	Endocrine Research Center, Institute of Endocrinology and Metabolism, Iran University of Medical Sciences, Tehran, Iran
Sepideh Nezhadi	Faculty of Pharmacy, Tehran University of Medical Sciences, Tehran, Iran
Somayeh Handali	Faculty of Pharmacy, Tehran University of Medical Sciences, Tehran, Iran
Souvik Chakraborty	Department of Pharmaceutics, JSS College of Pharmacy, JSS Academy of Higher Education & Research, Mysuru, Karnataka, India
Surajit Dey	Roseman University of Health Sciences, College of Pharmacy, Henderson, Nevada, USA
Vikas Jain	Department of Pharmaceutics, JSS College of Pharmacy, JSS Academy of Higher Education & Research, Mysuru, Karnataka, India

CHAPTER 1

The Story of Diabetes and its Causes

Ramin Malboosbaf[1,*] and **Neda Hatami**[1]

[1] *Endocrine Research Center, Institute of Endocrinology and Metabolism, Iran University of Medical Sciences, Tehran, Iran*

Abstract: Diabetes mellitus (DM) is a complex metabolic disorder whose rising prevalence is terrible. A deeper knowledge of the pathophysiology of diabetes could assist in discovering possible therapeutic targets for treating diabetes and its associated problems. The common feature of diabetes, regardless of the specific pathology involved, is hyperglycemia brought on by the death or dysfunction of β-cell. As insulin deficiency gets worse over time, dysglycemia progresses in a continuum. This chapter has provided a brief review of the pathophysiology of diabetes. Also, the roles of genetics and environmental factors have been emphasized.

Keywords: Diabetes, Disease, Factor, Glucose, Pathophysiology.

INTRODUCTION

Diabetes mellitus is a complex metabolic disorder whose principal clinical and diagnostic feature is hyperglycemia [1]. Diabetes has reached epidemic proportions; the global diabetes prevalence in 20-79-year-old in the latest reports was estimated to be 10.5% (536.6 million people), rising to 12.2% (783.2 million) in 2045 [2]. Over the next 20 years, its prevalence is expected to double, affecting more than half a billion people, with more than 75% of patients living in low- and middle-income countries [3]. Additionally, the increase in prevalence in developing countries is believed to be greater due to the widespread adoption of Western lifestyle habits, such as sedentary behavior, inactivity, and a high-energy diet [4, 5].

The risk of a variety of cardiovascular disorders is roughly doubled by diabetes, particularly type 2 diabetes mellitus (T2DM) [6]. In addition, a wide range of non-vascular diseases, such as cancer, infections, liver disease, and mental and nervous system disorders, are linked to T2DM [7]. In a similar vein, type 1 diabetes mellitus (T1DM) is linked to an increased risk of both vascular and non-

---

* **Corresponding author Ramin Malboosbaf:** Endocrine Research Center, Institute of Endocrinology and Metabolism, Iran University of Medical Sciences, Tehran, Iran; E-mail: malboosbaf.r@gmail.com

Ali Rastegari (Ed.)

vascular complications. A deeper knowledge of the pathophysiology of diabetes could assist in discovering possible therapeutic targets for treating diabetes and its associated problems [8, 9].

TYPE 1 DIABETES

The prevalence of T1DM is increasing worldwide. Although T1DM is often diagnosed in childhood, 84% of people living with T1DM are adults [10]; 62% of all new T1D cases in 2022 were in people aged 20 years or older [11]. T1DM affects men and women equally [12] and reduces life expectancy by an estimated 13 years [9]. With some exceptions, the incidence of T1DM is positively related to geographic distance north of the equator [13]. Colder seasons correlate with the diagnosis and progression of T1DM. Both disease onset and the incidence of islet autoimmunity appear to be higher in autumn and winter than in spring and summer [14 - 16].

Role of Genetics

The higher prevalence of T1DM in a family suggests a hereditary risk, which increases with the proband's degree of genetic similarity. Human leukocyte antigen (HLA) gene variations alter how the HLA protein binds to antigenic peptides and how the antigen is presented to T cells, contributing to 50-60% of the gene risk. Cell surface proteins involved in antigen presentation and self-tolerance are encoded by HLA genes, which are essential for controlling the immune response. As a result, genetic variations in these proteins' amino acid sequences may alter the repertoire of presented peptides and result in self-tolerance loss [17].

The autoimmune nature of diabetes is primarily due to its strong connection to HLA, the DQA and DQB genes, and its direct influence through the DRB genes [18]. Genome-wide association studies have demonstrated a strong link with the HLA-DR3 and HLA-DR4 haplotypes, as well as an exclusive link between the autoimmune destruction of β-cells and the DR4-DQB1I0302 haplotype [19 - 21].

Smaller effects are caused by about additional 50 genes individually [22, 23], including gene variants that modulate immune regulation and tolerance, viral responses [24 - 29], responses to environmental signals, and endocrine function [30]. Some variants are expressed in pancreatic β-cell [31]. In relatives, the onset and progression of islet autoimmunity are influenced by genetics [32, 33]. These gene variants collectively are responsible for 80% of T1DM inheritance [34]. A patient's risk, C-peptide decline rates, and response to various therapies can all be predicted by genetic variants [35]. With a deeper comprehension of heredity profiles, new goals for individualized interventions may be realized.

Role of the Environment

Numerous pieces of evidence suggest that environmental and genetic factors interact to cause autoimmunity and the development of T1DM, such as T1DM discordance rates in twins, the variance in geographic prevalence, and the adjustment of disease incidence rates as individuals migrate from low to high-incidence countries. The fact that most patients with the highest risk HLA haplotypes do not develop T1DM lends credence to this gene-environment interaction. Timing of environmental trigger exposure can also be very important. The investigation of environmental exposures is made more challenging by the variation in disease onset age. However, the early onset of islet autoantibodies linked to T1DM in children raises the possibility that early environmental exposures may play a role [10].

Infection

Congenital rubella infection has strong evidence to raise the possibility of T1DM development [36]. Enteroviruses are also thought to be associated with T1DM [37]. These infections are considered to alter gut microbiome composition [10].

Dietary Factors

β-cell autoimmunity can be affected by the timing of exposure to foods like grains and nutrients like gluten [10], as some studies show that early initiation of (<3 months) cereals may have this effect [38]. Retrospective studies led to the hypothesis that early initiation of cow's milk or less breastfeeding could increase the risk of T1DM. However, it was not confirmed by prospective studies [39]. Vitamin D deficiency and low levels of omega-3 fatty acids have been probably linked to an increased risk of T1DM [40].

Natural History and Prognosis

The common feature of diabetes, regardless of the specific pathology involved, is hyperglycemia brought on by the death or dysfunction of β-cell. As insulin deficiency gets worse over time, dysglycemia progresses in a continuum. The ability to categorize diseases and determine where and how to intervene best to stop or halt disease progression and complications depends on understanding the natural history of β-cell mass and function [10]. T1DM pathogenesis is influenced by both humoral and cellular immunity [41]. There is increasing evidence of significant overlap across the entire spectrum of diabetes, even though T1DM is caused by the immune system's destruction of beta cells, and T2DM is mostly associated with glucose-specific insulin secretion problems [42]. In both types of diabetes, the hyperglycemia-induced stress response may contribute to β-cell apoptosis [43]. The changes in β-cell phenotype that are the consequence of hyperglycemia may reflect the dedifferentiation of β-cell, which are important to the natural history and staging of diabetes [44].

Long before the diagnosis, abnormal insulin secretion can start [45 - 48], with a gradual decrease that starts at least two years before the diagnosis and accelerates immediately after diagnosis [49, 50]. In a similar time frame, a decrease in β-cell sensitivity seems to take place [51]. The late insulin response increases as the early insulin response subsides, suggesting a potential compensatory mechanism [52]. It has been said that the a decrease in insulin secretion in the first year after diagnosis is biphasic, steeper in the first year than in the second. Additionally, the data imply that adults experience a slower rate of decrease [53]. Until there is little to no insulin secretion, the loss of insulin secretion can persist for years after diagnosis. Though, even after 30 years of T1DM, the majority of patients still have low C-peptide levels [54]. Usually, glucose levels are high years before T1DM is diagnosed. Higher glucose levels, even within the normal range, signify T1DM [54 - 57]. There are significant fluctuations in glucose during the progression to T1DM [58]. Metabolic progression markers could be used to predict more accurately when people at risk will develop diabetes [35, 59]. Prediction can be further enhanced by combining dynamic glucose and C-peptide changes into risk scores [60, 61].

Diabetic ketoacidosis (DKA), as the first onset of the disease, can occur when there is sudden β-cell death in children and adolescents. In some, the course of the illness is prolonged, with a slight increase in fasting blood glucose that only becomes severe with or without ketoacidosis in the presence of physiological stresses like severe infections. Patients with this form of diabetes, despite the variable course, require insulin treatment for survival when they develop severe or complete insulin deficiency in early, middle, or even late life. Low or undetectable plasma C-peptide levels are a sign of severe or complete insulin deficiency, regardless of the age of onset [62 - 64]. Fig. (**1**) depicts the natural history of T1DM [64].

Before a clinical diagnosis of T1DM is made, circulating autoantibodies against insulin, glutamic acid decarboxylase (GAD), the protein tyrosine phosphatase IA-2, and the zinc transporter eight can be found [65]. Reversion is uncommon in people with multiple autoantibodies, while individuals with single autoantibodies often become negative [66]. Anti-GAD-65 is the most important, detectable in approximately 80% of patients at the time of clinical diagnosis, followed by Islet cell antibodies (ICAs) and IA-2, which are present in 69-90% and 54-75%, respectively. In T1DM, Anti-GAD-65 gradually decreases over time. In high-risk populations, the presence of anti-GAD antibodies is a strong predictor of T1DM development in the future [67]. About 70% of all infants and young children at the diagnosis have insulin autoantibodies (IAAs), which also play a significant inhibitory role in insulin function in patients receiving insulin therapy [18].

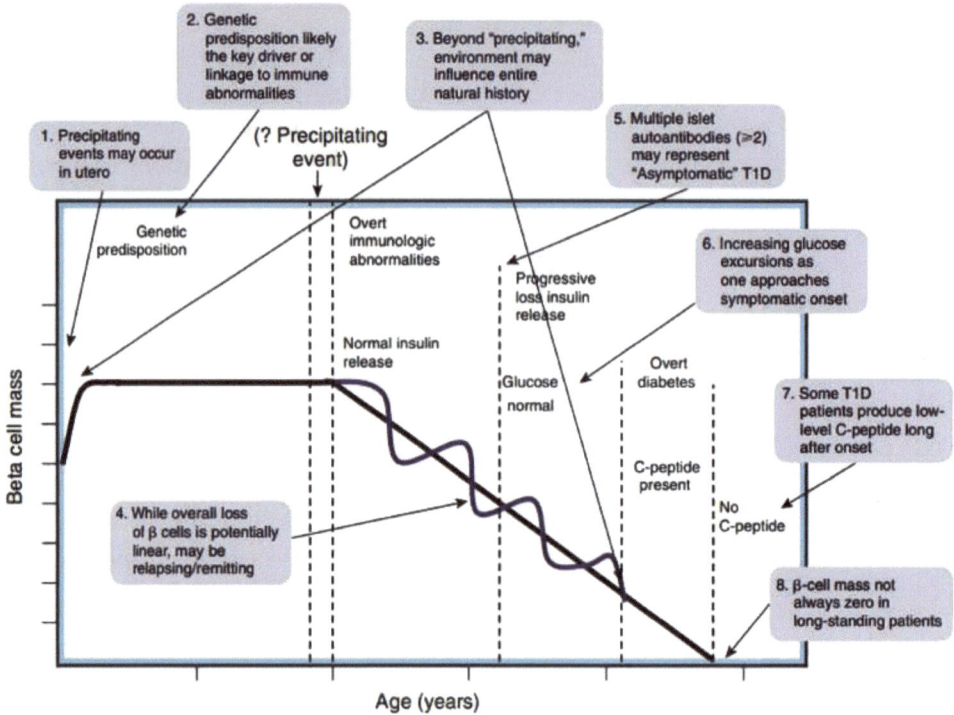

Fig. (1). Pathophysiological mechanisms in common for NADLD and T2DM. LPS: lipopolysaccharides; CRP: C reactive protein; TNF-α: tumor necrosis factor; IL-6: interleukin-6; ROS: reactive oxygen species; TLR: toll-like receptor [119].

Children with HLA risk genotypes in relatives with T1DM who have two or more islet autoantibodies have a 75% chance of developing clinical diabetes in the next ten years [68]. As an increasing number of autoantibodies are found, the risk increases [68 - 70]. Today, a diagnostic stage of T1DM is thought to be a positive test for at least two autoantibodies [35]. Although autoimmunity in T1DM has a significant prognostic value, there is no effective treatment or prevention strategy [71].

It is interesting to note that 5% of T2DM patients have autoantibodies to GAD. These patients have a lower BMI and less residual β-cell function than patients with T2DM who do not have a GAD antibody. Additionally, they have a genetic profile that is more in line with that of T1DM patients and a previous need for insulin therapy, indicating that adult autoimmune diabetes may be a form of T1DM with a slower course and a later onset age [72].

Effects on Glucose Metabolism

While insulin insufficiency is the main flaw in T1DM, insulin operation is also flawed. The expression of multiple genes is required for target tissues to correctly respond to insulin, including glucokinase in the liver and the GLUT-4 class of glucose transporters in adipose tissue. Lack of insulin causes unregulated lipolysis and high plasma-free fatty acid levels, which inhibit the metabolism of glucose in peripheral tissues like the skeletal muscle [73].

Hepatic glucose output is increased in uncontrolled T1DM at first through glycogenolysis, followed by gluconeogenesis. Insulin also controls hepatic glucokinase. So, decreased glucose phosphorylation induces a higher glucose transport into the blood [73]. Additionally, non-hepatic tissues' glucose utilization is affected by insulin deficiency. Adipose tissue and skeletal muscle are particularly affected by insulin's effect on glucose uptake. This is done by transporting glucose transporter proteins to the plasma membrane of these organs, which is mediated by insulin [73].

Effect on Lipid Metabolism

After eating, the main function of insulin is to promote the storage of dietary energy in the form of glycogen in hepatocytes and skeletal muscle. Additionally, insulin causes hepatocytes to produce triglycerides and store them in adipose tissue. Normally, insulin is required for lipoprotein lipase (LPL) to act on plasma triglycerides. Fatty acids can be extracted from circulating triglycerides and stored in adipocytes by LPL, a membrane-bound enzyme on the surface of the endothelial cells that line the vessels. The lack of insulin leads to hypertriglyceridemia [73]. Triglycerides quickly mobilize in uncontrolled T1DM, raising plasma-free fatty acid concentrations. Except for the brain, many tissues absorb the free fatty acids and convert them to produce energy [73].

Malonyl-COA levels decrease, and fatty acetyl-COA transport into mitochondria rises in the absence of insulin. Acetyl-COA is produced when fatty acids are oxidized by mitochondria, which can be further oxidized during the tricarboxylic acid (TCA) cycle. Most of the acetyl-COA found in hepatocytes is metabolized into the ketone bodies rather than being oxidized by the TCA cycle. The brain, heart, and skeletal muscles use these ketone bodies for energy [73].

The decreased utilization of glucose is exacerbated in T1DM by the increased availability of free fatty acids and ketone bodies. Ketoacidosis occurs when the body produces more ketone bodies than it can use. Acetone, exhaled from the lungs, provides a characteristic odor in breath, which is a spontaneous breakdown product of acetoacetate.

Effects on Protein Metabolism

By accelerating protein synthesis and decelerating protein breakdown, insulin has an overall positive impact on protein metabolism. As a result, low insulin levels lead to more protein breakdown. The plasma's concentration of amino acids rises as a result of the enhanced rate of proteolysis. Hepatic and renal gluconeogenesis, which furthers hyperglycemia, is facilitated by using glucogenic amino acids as precursors [73].

PATHOPHYSIOLOGY OF TYPE 2 DIABETES

The Centers for Disease Control and Prevention (CDC) recently released the 2022 National Diabetes Statistics Report. This report estimates that more than 130 million adults in the United States live with diabetes or are prediabetic. Diabetes data with respect to income level was published for the first time, showing that a higher prevalence of diabetes was also associated with poverty. People with less education were more likely to be diagnosed with diabetes [74].

Patients at risk of T2DM (obese patients and first-degree relatives) show an initial state of insulin resistance, which is compensated by hypersecretion of insulin by β-cell (hyperinsulinemia). However, this functional reserve of the pancreas will eventually be unable to produce the necessary amounts of insulin. Obese euglycemic people have 30% less insulin sensitivity than lean euglycemic people, and they secrete more insulin to keep their normal glucose tolerance (euglycaemic hyperinsulinemia). An elevated blood glucose concentration (hyperglycemic hyperinsulinemia) occurs over time when obese euglycemic individuals lose their ability to compensate for hyperinsulinemia while experiencing more insulin insensitivity (hyperglycemic hyperinsulinemia). When diabetes is diagnosed, apparent hyperglycemia (hyperglycemic hypoinsulinemia) occurs when beta cells are unable to produce sufficient insulin [75]. It is generally acknowledged that abnormal insulin sensitivity may occur up to 15 years before a clinical diagnosis of diabetes is made, even though the proportions of β-cell dysfunction and insulin resistance's contributions may differ. These pathophysiological studies' results illustrate compensatory hyperinsulinemia's failure as a hallmark of overt hyperglycemia. Therefore, in addition to mechanistic studies investigating underlying mechanisms of hyperglycemia, more recent study has concentrated on the routes leading to a β-cell failure [76].

Insulin resistance can be measured using a variety of techniques. The gold standard for determining insulin sensitivity/resistance in an individual is a hyperinsulinemic-euglycemic clamp. In this technique, developed by DeFronzo, a patient is given a constant insulin infusion to create hyperinsulinemia. A second infusion containing glucose is administered concurrently and adjusted to produce euglycemia. Since the subject is in a steady state, the glucose infusion rate represents the glucose uptake/release rate in muscle, fat, and other tissues under hyperinsulinemic conditions (*i.e.*, the patient's insulin sensitivity) [77]. To assess insulin resistance, HOMA-IR (homeostatic model assessment of insulin resistance) can also be calculated using only fasting glucose and insulin levels [78]. Insulin sensitivity is influenced by several factors, including age, weight, race, body fat (particularly intra-abdominal obesity), physical activity, food intake, gut microbiota, and medications. Extensive studies show that the development of impaired glucose tolerance (IGT) and diabetes is significantly influenced by insulin resistance. Environmental factors also significantly impact a person's genetic predisposition to insulin resistance and, consequently, diabetes [79 - 82].

Insulin-sensitive Tissues

Although much of our understanding of insulin's action on metabolism concerns the insulin-sensitive metabolic tissues of muscle, liver, and fat, insulin receptors are ubiquitously expressed, and the action of insulin on non-classical tissues is crucial to health (Fig. **2**). Insulin signaling in β-cell is critical for these cells' adaptive survival and proliferation. Chronic mild hyperglycemia increases insulin secretion and β-cell mass, and the ability to upregulate β-cell mass affects the development of T2DM. Interestingly, when insulin receptors are removed in the β-cell first phase, insulin secretion is lost, and compensatory β-cell growth is impaired in response to dietary obesity and hepatic insulin resistance [83, 84], indicating an unexpected autocrine loop. In addition, FoxO transcription factors, downstream targets of insulin action, are important in maintaining β-cell function and identity [85], and β-cell seems to dedifferentiate when FoxOs are deleted [86]. Evidence that β-cell dedifferentiation into insulin-resistant conditions is emerging in humans [87], but the extent to which this phenomenon accounts for β-cell failure in T2DM and whether this process can be manipulated are still under investigation.

The action of insulin on glucagon-producing alpha cells suppresses glucagon secretion and maintains glucose homeostasis. This has been proven by the deletion of insulin receptors on mouse alpha cells, resulting in mild glucose intolerance, increased glucagon secretion, and progressive hyperinsulinemia [88]. Therefore, insulin signaling to alpha and beta cells in islets is important for metabolic health and glucose homeostasis.

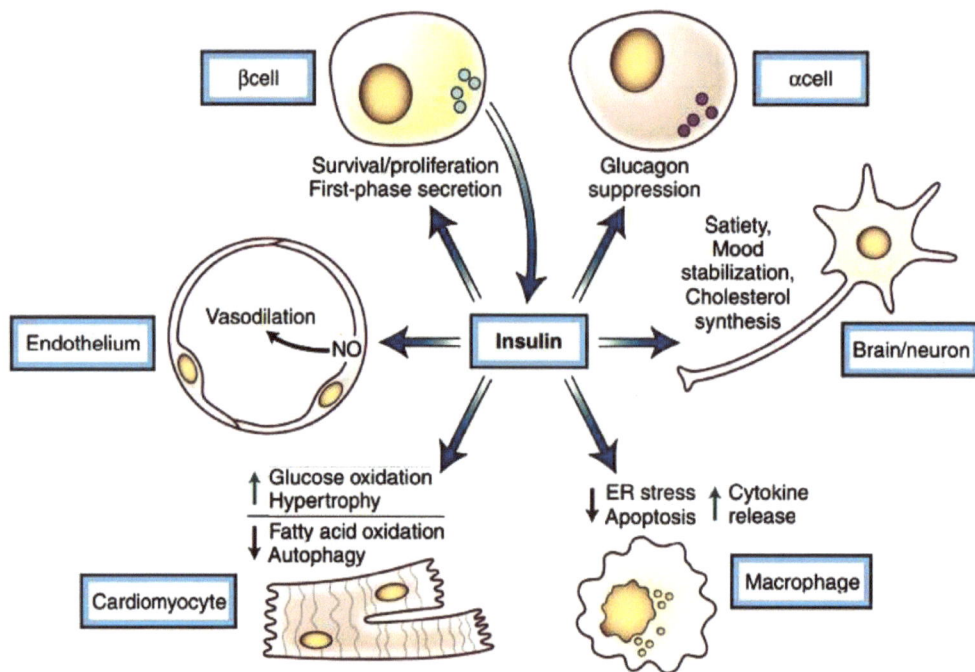

Fig. (2). Insulin-sensitive tissues. Insulin receptors are expressed in many "nonclassical" tissues. ER, endoplasm mic reticulum. Adapted from Williams [100].

Diabetes-related vascular complications are a significant cause of morbidity and mortality. Therefore, it is unsurprising that vascular endothelium and cardiomyocytes are excellent insulin stimulators. In endothelial cells, insulin action promotes vasodilation *via* nitric oxide production, which is diminished in insulin-resistant states [89 - 91]. Endothelial insulin sensitivity is also important for the transendothelial transport of insulin to peripheral tissues. The relative permeability of the endothelial barrier in different tissues plays a crucial function in the timing of insulin action in these tissues after a bolus of insulin [92]. Atherosclerosis is also significantly influenced by endothelial insulin signaling. In an atherogenic animal model, insulin receptor knockout in endothelial cells led to a two- to three-fold rise in atherosclerotic lesions [93]. Conversely, enhancing downstream activation of insulin signaling either through overexpression of IRS or inhibition of FoxOs improves vascular endothelial function and may prevent atherosclerosis [94, 95].

The insulin signaling cascade in cardiomyocytes controls growth and metabolism under physiological and pathological conditions [96, 97]. The heart has a remarkable ability to utilize a variety of substrates for energy production, but in the fasted state, fatty acids are preferentially metabolized. Upon insulin stimulation, glucose oxidation is increased, and fatty acid oxidation is suppressed, but this metabolic flexibility is lost in insulin-resistant or diabetic states [98]. Insulin may also affect cardiovascular risk through its effects on immune cells, particularly macrophages. Macrophages play an important role in atherosclerosis development by forming foam cells in atherosclerotic plaques, increasing inflammation and apoptosis, resulting in a necrotic core prone to rupture [99, 100].

Insulin Resistance and the Risk of Type 2 Diabetes Mellitus

Since many years ago, obesity and T2DM have been linked. Insulin resistance is more strongly associated with central (also known as intra-abdominal or visceral) obesity than total obesity [101 - 107]. Furthermore, some research indicates that subcutaneous fat could prevent insulin resistance [108]. Subcutaneous fat is less lipolytically active than abdominal fat, possibly due to the latter's higher number of adrenergic receptors [109]. Additionally, abdominal fat stores are resistant to insulin's antilipolytic effects, which include changes in lipoprotein lipase activity. This causes an increase in lipase activity and a greater flux of fatty acids into the bloodstream, with portal circulation receiving the highest fatty acid load [110].

Conversely, subcutaneous fat produces and releases more adiponectin, which is a beneficial adipokine. Moreover, intra-abdominal fat's high levels of 11-hydroxysteroid dehydrogenase type 1 (HSD11B1) may increase the conversion of inactive cortisol into active cortisol, thereby boosting local cortisol production. This may change adipocytes to promote lipolysis and change the synthesis of adipokines, which can directly influence glucose metabolism [110].

Obesity affects the development of T2DM on genetic background, making it more than merely a risk factor. A series of pathophysiological events cause obesity to progress into T2DM:

(a) Increase in adipose tissue mass, leading to increased lipid oxidation;

(b) Insulin resistance detected early in obesity, manifested by euglycemic bracketing as resistance to insulin-mediated glucose storage and oxidation, thereby blocking glycogen cycle function;

(c) Unused glycogen prevents further glucose storage despite sustained insulin secretion, leading to T2DM;

(d) Complete depletion of β-cell occurs later [67].

It has long been recognized that the liver and muscles are the main sources of systemic insulin resistance [111]. During a fast, the liver generates glucose from non-glucose substrates (gluconeogenesis) to guarantee the constant availability of a carbohydrate energy source [112]. Numerous investigations have shown that elevated gluconeogenesis exists in T2DM patients, even in hyperinsulinemia, indicating that hepatic insulin resistance is a major factor in the development of fasting hyperglycemia [113, 114]. Although the causes of decreased hepatic insulin sensitivity are unclear, liver fat accumulation (steatosis) is thought to play a significant role [115]. It is interesting to note that hepatic steatosis occurs before overt T2DM and is frequently linked to obesity [116], especially visceral (or abdominal) obesity [117].

It is generally acknowledged that an accumulation of fat in the subcutaneous tissue results from a permanently positive energy balance brought on by too many calories and insufficient exercise. Fat is transferred to ectopic compartments like the liver, pancreas, muscle, perivascular, pericardium, and omentum when this limit is met (the spill-over or adipose tissue overflow hypothesis) [117]. Because insulin signaling is disrupted intracellularly, liver and muscle fat accumulation impairs insulin-mediated glucose absorption [111]. Notably, lean T2DM has been demonstrated to have muscle insulin resistance, indicating that body fat-independent mechanisms may also play a role in the etiology of this condition [118]. On the other hand, fat accumulation within the pancreatic islets determines β-cell malfunction and an increase in plasma glucose, which reduces the insulin response to ingested glucose (twin-cycle hypothesis) [116].

Non-alcoholic fatty liver disease(NAFLD) and T2DM are two diseases with the same entity [119]. The prevalence of NAFLD among people with T2DM is too high [120] which is estimated to be more than 70% [121, 122]. Lately, in a meta-analysis, NAFLD people were shown to be more susceptible to developing T2DM with even more risk in "severe" NAFLD [123]. The pathophysiological connection between these two diseases seems to be insulin resistance (IR) [124]. Moreover, their close association is shown to cause diabetic patients with NAFLD to encounter worse outcomes because of more probability of liver cirrhosis, HCC, and death [125, 126]. NAFLD has also been considered to cause more risk of side effects related to the kidneys and the heart [127, 128]. The detailed pathophysiologic mechanism connecting NAFLD and T2DM is not a matter of debate in this chapter, so a schematic summary is illustrated in Fig. (**1**).

Adipose Tissue and Insulin Resistance

To maintain metabolic homeostasis and increase cell mass, excess calories must be used or stored when nutrient intake surpasses energy expenditure. Most excess nutrients, whether carbohydrates, proteins, or lipids, are ultimately deposited as triglycerides in white adipose tissue. Lipids and other nutrients enter non-storage tissue (myocytes, hepatocytes, vascular cells, and β-cells) when the storage capacity of adipose tissue is surpassed. It can produce toxic lipid metabolites that trigger the activation of protein kinase C (PKC) isoforms, resulting in insulin resistance [129].

Some research indicates that insulin resistance may be exacerbated by an inability to increase fat mass in response to overeating [130]. Storage cells are only one aspect of adipocytes. They modulate the body's fatty acid uptake and release, take part in the glycerol fatty acid cycle, release leptin and other hormones that indicate the body's energy status, and emit an increasing number of cytokines that have hormonal, paracrine, and autocrine effects [131].

The accumulation of excess nutrients can hurt the adipocytes themselves, causing events that can negatively affect the body. The expression of leptin, interleukin-6 (IL-6), IL-8, monocyte chemoattractant protein 1 (MCP1), and granulocyte colony-stimulating factor (G-CSF) is all raised with increasing adipocyte surface area in obesity. Pro-inflammatory macrophages of the M1 type are drawn to these and possibly other cytokines. These macrophages release factors like tumor necrosis factor (TNF), which can have local and systemic inflammatory effects [132].

In addition to energy-storing white adipose tissue, mammals like humans also have brown adipose tissue that burns energy. Brown adipose tissue is a thermogenic tissue that contains unique adipocytes with diverse morphology, including multiloculated lipid droplets, increased mitochondrial content, and expression of uncoupling protein 1 (UCP1) to uncouple electron transport and generate heat. The sympathetic nervous system stimulates brown adipose tissue, increasing fatty acid mobilization and oxidation. Age and obesity are negatively correlated with brown adipose tissue content, which may be a consequence of or a contributor to insulin resistance in these conditions. Brown adipose tissue in the neck and supraclavicular regions increases mass and activity when humans are exposed to cold for an extended or repeated time, defined by glucose uptake, and improves glucose homeostasis.

Role of Gut Microbiome and Metabolome in Diabetes and Insulin Resistance

There is increasing evidence that the gut microbiome, more specifically, the bacteria that live in the gastrointestinal tract, plays a significant role in how diet and other environmental factors affect diabetes [133]. In mammals, the gut microbiome is seeded at birth; it is reformed during childhood and then becomes relatively stable in adulthood, although diet, antibiotics, and a variety of disease states can reshape it. Most gut microbiotas are considered commensal (not harmful) or mutual (beneficial) to the host. In addition to preserving the integrity of the intestinal mucosa, these bacteria also serve critical roles in immunological regulation, pathogen defense, and the metabolism of nutrients, foreign substances, and medications. These basic functions affect many aspects of normal physiological function. In the last decade, it has become clear that the gut microbiome possibly exacerbates obesity, diabetes, metabolic syndrome, and insulin resistance in rodents [134] as well as in humans [135] and acts as an integrator and mediator of some of the effects of genetics, nutrition, and bariatric surgery. It has been demonstrated in mice that low-dose antibiotic therapy at a young age can increase the risk of obesity and glucose intolerance by disrupting the establishment of a normal microbiome [136].

The function of the gut barrier, digestion of normally indigestible dietary components, bile acid modification, gut development, and immune system development are all significantly impacted by the gut microbiota [133, 137]. As a result of these impacts, bacterial proteins, endotoxins, and cytokines may be released into the bloodstream [138, 139], and hundreds of metabolites, such as bile acids, short-chain fatty acids, amino acids, and many other types of compounds, may change [140, 141]. Collectively, these cause immune activation and tissue-specific metabolic dysregulation, which in turn causes insulin resistance and the development of diabetic pathogenesis. In rodents and humans, several metabolites have been demonstrated to be associated with insulin resistance; some of these including 2-aminoadipate, β-hydroxybutyrate, and N-acetyl glycine, are thought to be biomarkers of type 2 diabetes [142]. Further studies will be required to accurately understand how the gut microbiota influences insulin sensitivity and diabetes progression, as well as if therapies that change the gut microbiota may be utilized to treat or prevent T2DM. In this regard, several options are possible, ranging from the administration of prebiotics (nutrients that alter the microbiota's composition in the gut) and probiotics (mixtures of bacteria themselves) to the transplantation of a healthy microbiome through fecal transfer.

Role of Gut Hormones

In the last two to three decades, it has been discovered that gut hormones play a significant role as one-cell regulators. When glucose is consumed orally, insulin secretion is higher than when glucose is infused intravenously with a superimposed glucose excursion (glycemic infusion). This finding implies that after consuming glucose, there are substances that trigger the release of insulin. These elements are called incretins, gastrointestinal messengers that can increase insulin secretion [143].

The primary hormones causing this phenomenon have been identified as two incretins: glucagon-like peptide-1 (GLP-1), which is secreted by L-cells primarily found in the ileum and colon, and gastric inhibitory polypeptide (GIP), which is secreted by enteroendocrine K-cells primarily found in the duodenum and proximal jejunum. These polypeptides released after eating can boost insulin (GLP-1 and GIP) secretion and decrease glucagon (GLP-1) secretion [143]. Although there does not seem to be a significant secretory defect in GIP secretion in T2DM, significantly decreased GLP-1 secretion has consistently been identified in T2DM, insulin resistance, and obesity [143 - 145]. As a result, patients with T2DM experience an attenuated GLP-1 response after food intake, which leads to reduced postprandial levels of both GLP-1 and insulin as well as relative hyperglucagonemia. In addition, T2DM patients are less responsive to GIP and GLP-1, which can be ameliorated by restoring euglycemia. This suggests that the loss of incretin is not the direct cause of hyperglycemia but rather a side effect [146].

It has long been understood that the kidney regulates blood glucose [147]. The glomerulus filters approximately 180 g of glucose daily: The membrane transporter sodium-glucose cotransporter 2 (SGLT2) resorbs approximately 90% in the proximal tubule and 10% in the descending tubule (SGLT1) [148]. Until the maximum reabsorption capacity (Tm), which is 11.0 mmol/L (200 mg/dL) in healthy adults, is exceeded, the rate of filtered glucose reabsorption increases linearly [149]. An important finding is that Tm levels seem to be higher in T2DM patients, exacerbating hyperglycemia and creating a vicious cycle [150]. Upregulation of SGLT2 as a result of hyperglycemia is the mechanism connected to the change in Tm [151].

Genetics

Although a subset of genetic variants are linked to type 1 and type 2 diabetes [152, 153], the genetic basis for the two diseases is vastly distinct. It could be used to classify diabetes [154]. More than 130 genetic variants linked to T2DM, glucose levels, or insulin levels have been discovered by genome-wide association studies; nevertheless, these variants account for less than 15% of the disease's heritability [76, 155]. There are many possibilities to explain most of the heritability of T2DM, such as disease heterogeneity, gene interactions, and epigenetics. The majority of type 2 variants can be found in non-coding genomic areas. The parent of origin significantly impacts some variants, such as KCNQ1 [112]. Children born to moms carrying KCNQ1 may have less functional β-cell mass at birth and, as a result, may be less able to enhance their insulin secretion when faced with insulin resistance [113].

The search for rare variants that protect against T2DM, such as Loss-of-function mutations in SLC30A8, which could offer potential new drug targets for T2DM, was another topic of great interest [114]. So far, however, there has been little improvement in the predictive value of known genetic variants compared to conventional clinical risk factors (BMI, family history, and glucose) for T2DM. Although, at this point, genetics have not been successful in further differentiating subclasses of diabetes, aside from the profiles that differentiate between T1DM and T2DM and a small number of specific variants that identify small subgroups of patients (MODY), due to the rapid development of molecular genetic tools and declining costs for next-generation sequencing [10].

Genetically predisposed individuals eventually develop dysfunctional β-cells because of the higher insulin production and secretion requirement. However, of the suggested potential processes producing β-cell dysfunction, their comparative and chronological roles are unknown. The balance between alpha and beta cell mass and function inside the islets of Langerhans may be altered by the stressed β-cell, according to certain theories. Remarkably, insulin limits glucagon release by exerting a negative paracrine effect on α-cells; Consequently, a shortage of insulin causes glucagon levels to rise, which then triggers hepatic gluconeogenesis to raise blood glucose levels even further.

In contrast to T1DM, this disease has not been linked to genes involved in the immune response, including autoimmunity. As a result, pancreatic cells are not destroyed by the immune system [157, 158].

Environmental Influences

When β-cells cannot release enough insulin to meet demand, typically associated with increased insulin resistance, T2DM develops. Islet autoimmunity is seen in a small percentage of T2DM patients [159, 160]. Obesity, which has a complicated genetic and environmental etiology, is a major risk factor for T2DM [161, 162]. Ectopic fat deposition in the liver and muscle leads to the development of insulin resistance. Fat can accumulate in the pancreas as well, which can lead to islet inflammation, a loss in β-cell function, and eventually β-cell death [163].

T2DM occurs in people with different BMI/body fat compositions in Asians and Asian Americans with lower BMI [164]. In susceptible individuals, there may be a personal fat threshold in vulnerable individuals, above which ectopic fat accumulation worsens insulin resistance and results in β-cell decompensation. Losing weight may reduce the accumulation of fat in the pancreas and improve insulin sensitivity in the liver and skeletal muscle [165, 166]. In prediabetes and newly diagnosed T2DM, energy restriction and weight loss can at least partially reverse insulin secretion defects [167]. Unfortunately, even with the significant weight loss brought on by bariatric surgery, longstanding diabetes is difficult to reverse.

Sleeping

There are significant human epidemiological associations between sleep deprivation and increased rates of obesity and other metabolic disorders, including T2DM [168]. Experimental sleep disorders may impact other cardiovascular risk variables, directly affect insulin action, change leptin and ghrelin release resulting in appetite stimulation, and enhance the production of inflammatory cytokines. The relationship between the appearance of nutrients and enzymes that break down nutrients can be altered by altering normal dietary patterns tuned into circadian metabolism. Changes in the appearance of fatty acids and the activity of lipoprotein lipase, for instance, can alter the distribution of lipids to sensitive tissues, resulting in lipotoxicity, decreased leptin secretion, and increased appetite [169]. Additionally, obstructive sleep apnea, which causes both hypoxemia and fragmented sleep, increases the risk of developing diabetes and insulin resistance. There is increasing evidence that treating sleep apnea can improve glucose control in T2DM patients, although treatment adherence is the principal barrier to success [170].

Natural History and Prognosis

The pathophysiology of T2DM is mostly caused by defective insulin secretion. Insulin production fluctuates greatly in response to insulin sensitivity to maintain normal glucose levels. Insulin sensitivity and secretion have a curvilinear connection represented by a disposition index [171]. A low disposition index and inability to overcome insulin resistance characterize people with type 2 diabetes. So, although absolute insulin levels in obese T2DM patients with insulin resistance may be higher than those in insulin-sensitive lean controls, these levels are still below what is necessary to account for their insulin resistance.

First-phase insulin secretion is significantly reduced or eliminated, particularly in response to glucose stimulation [172]. The proinsulin to insulin (C-peptide) ratio is high in T2DM, and maximum insulin secretion and potentiation of insulin responses to non-glucose stimuli are greatly diminished [172, 173]. Hyperglycemia usually gets worse over time and is harder to treat. The continuing decline in β-cell function is typically the cause of T2DM's progressive character. Dysglycemia is a continuum that progresses from normal to overt diabetes, while absolute thresholds are used to diagnose prediabetes and diabetes [174].

A window for treatment that can prevent or delay the progression of the disease and its complications is provided by early screening [175, 176]. At this time, the majority of physicians do not fully manage these patients, and patients with overt diabetes face years of hyperglycemia because the intensification of treatment is frequently postponed. Numerous studies have demonstrated that adopting a healthier lifestyle or taking medication can slow the progression of prediabetes to diabetes. In addition, early treatment has been shown to reduce retinopathy and cardiovascular and all-cause mortality. According to the evidence, early detection of prediabetes and maintaining blood glucose levels close to normal may change the course of the disease's natural history [177 - 179].

Maturity-onset Diabetes of the Young, Latent Autoimmune Diabetes of Adults, and Double Diabetes

With a better understanding of diabetes pathophysiology, it has also become clearer that some forms of diabetes do not fall fully into the Type 1 or Type 2 categories. The term "Maturity-onset Diabetes of the Young" (MODY) describes single-gene diseases that cause T2DM-like conditions in younger age groups. These monogenic forms, which may account for 2% of all diabetes cases in the UK [180], show an autosomal dominant inheritance pattern of diabetes and are usually diagnosed in childhood or adolescence. While the natural history is variable, it is typically gradual to go from normal to mild hyperglycemia and overt diabetes [181].

In the middle of the 1980s, the first genetic studies of MODY began, and several genetic defects were found. MODY2 (mutation of the glucokinase (GCK) gene) and MODY3 (hepatocyte nuclear factor) are the two most common disorders. MODY3 has a remarkable response to sulfonylureas, which are drugs that stimulate insulin secretion. Specifically, pancreatic cells' ineffective insulin production and release are characteristic of recognized MODY forms [182].

A type of diabetes known as latent autoimmune diabetes of adults (LADA), also referred to as slowly progressive insulin-dependent diabetes, is characterized by three characteristics: adulthood at the time of diagnosis, autoantibodies associated with diabetes, and no need for insulin therapy at the time of diagnosis [183]. LADA accounts for 2–12% of all adult-onset diabetes, making it the most prevalent kind of autoimmune diabetes. GADAs, followed by ICAs, are the most significant and sensitive autoantibodies for LADA; others are only detectable in a few cases in LADA [184]. From a pathophysiological point of view, LADA could be regarded as slowly progressing T1DM.

In addition, incidences of combined T1DM and insulin resistance are rising with the rising prevalence of obesity and insulin resistance, leading to the definition of type 1.5 [185] or double diabetes [186]. Mostly, when some β-cell insulin is initially conserved, these patients present a diagnostic challenge.

CONCLUSION

Diabetes mellitus is a complex metabolic disorder whose principal clinical and diagnostic feature is hyperglycemia. In this disease, as insulin deficiency gets worse over time, dysglycemia progresses in a continuum. Studies have shown genetic and environmental factors have the main role in the development of this disease.

REFERENCES

[1]   Zaccardi F, Webb DR, Yates T, Davies MJ. Pathophysiology of type 1 and type 2 diabetes mellitus: A 90-year perspective. Postgrad Med J 2016; 92(1084): 63-9.
        [http://dx.doi.org/10.1136/postgradmedj-2015-133281] [PMID: 26621825]

[2]   Sun H, Saeedi P, Karuranga S, *et al.* IDF Diabetes Atlas: Global, regional and country-level diabetes prevalence estimates for 2021 and projections for 2045. Diabetes Res Clin Pract 2022; 183(109119)109119.
        [http://dx.doi.org/10.1016/j.diabres.2021.109119] [PMID: 34879977]

[3]   International Diabetes Federation . IDF diabe atlas. Brussels International Diabetes Federation 2015.

[4]   Shaw JE, Sicree RA, Zimmet PZ. Global estimates of the prevalence of diabetes for 2010 and 2030. Diabetes Res Clin Pract 2010; 87(1): 4-14.
        [http://dx.doi.org/10.1016/j.diabres.2009.10.007] [PMID: 19896746]

[5]   Chan JCN, Malik V, Jia W, *et al.* Diabetes in Asia. JAMA 2009; 301(20): 2129-40.
        [http://dx.doi.org/10.1001/jama.2009.726] [PMID: 19470990]

[6]   Sarwar N, Gao P, Seshasai SR, *et al.* The emerging risk factors collaboration diabetes mellitus, fasting blood glucose concentration, and risk of vascular disease: A collaborative meta-analysis of 102 prospective studies. Lancet 2010; 375(9733): 2215-22.
        [http://dx.doi.org/10.1016/S0140-6736(10)60484-9] [PMID: 20609967]

[7]   Rao Kondapally Seshasai S, Kaptoge S, Thompson A, *et al.* Emerging risk factors collaboration diabetes mellitus, fasting glucose, and risk of cause-specific death. N Engl J Med 2011; 364(9): 829-41.
        [http://dx.doi.org/10.1056/NEJMoa1008862] [PMID: 21366474]

[8]   Lind M, Svensson AM, Kosiborod M, *et al.* Glycemic control and excess mortality in type 1 diabetes. N Engl J Med 2014; 371(21): 1972-82.
        [http://dx.doi.org/10.1056/NEJMoa1408214] [PMID: 25409370]

[9]   Livingstone SJ, Levin D, Looker HC, *et al.* Scottish diabetes research network epidemiology group scottish renal registry estimated life expectancy in a scottish cohort with type 1 diabetes, 2008-2010. JAMA 2015; 313(1): 37-44.
        [http://dx.doi.org/10.1001/jama.2014.16425] [PMID: 25562264]

[10]   Skyler JS, Bakris GL, Bonifacio E, *et al.* Differentiation of diabetes by pathophysiology, natural history, and prognosis. Diabetes 2017; 66(2): 241-55.
        [http://dx.doi.org/10.2337/db16-0806] [PMID: 27980006]

[11]   International Diabetes Federation IDF Diabetes Atlas. 10th ed., Brussels, Belgium: International Diabetes Federation 2021.

[12]   Maahs DM, West NA, Lawrence JM, Mayer-Davis EJ. Epidemiology of type 1 diabetes. Endocrinol Metab Clin North Am 2010; 39(3): 481-97.
        [http://dx.doi.org/10.1016/j.ecl.2010.05.011] [PMID: 20723815]

[13]   Karvonen M, Viik-Kajander M, Moltchanova E, Libman I, LaPorte R, Tuomilehto J. Incidence of childhood type 1 diabetes worldwide. Diabetes Mondiale (DiaMond) Project Group. Diabetes Care

2000; 23(10): 1516-26.
[http://dx.doi.org/10.2337/diacare.23.10.1516] [PMID: 11023146]

[14]   Knip M, Veijola R, Virtanen SM, Hyöty H, Vaarala O, Åkerblom HK. Environmental triggers and determinants of type 1 diabetes. Diabetes 2005; 54 (2): S125-36.
[http://dx.doi.org/10.2337/diabetes.54.suppl_2.S125] [PMID: 16306330]

[15]   Noble JA, Valdes AM, Varney MD, *et al.* Type 1 diabetes genetics consortium hla class i and genetic susceptibility to type 1 diabetes: results from the type 1 diabetes genetics consortium. Diabetes 2010; 59(11): 2972-9.
[http://dx.doi.org/10.2337/db10-0699] [PMID: 20798335]

[16]   Hu X, Deutsch AJ, Lenz TL, *et al.* Additive and interaction effects at three amino acid positions in hla-dq and hla-dr molecules drive type 1 diabetes risk. Nat Genet 2015; 47(8): 898-905.
[http://dx.doi.org/10.1038/ng.3353] [PMID: 26168013]

[17]   Ziegler AG, Bonifacio E. Babydiab-babydiet study group age-related islet autoantibody incidence in offspring of patients with type 1 diabetes. Diabetologia 2012; 55(7): 1937-43.
[http://dx.doi.org/10.1007/s00125-012-2472-x] [PMID: 22289814]

[18]   Banday MZ, Sameer AS, Nissar S. Pathophysiology of diabetes: An overview. Avicenna J Med 2020; 10(4): 174-88.
[http://dx.doi.org/10.4103/ajm.ajm_53_20] [PMID: 33437689]

[19]   Burton PR, Clayton DG, Cardon LR, *et al.* Wellcome trust case control consortium genome-wide association study of 14,000 cases of seven common diseases and 3,000 shared controls. Nature 2007; 447(7145): 661-78.
[http://dx.doi.org/10.1038/nature05911] [PMID: 17554300]

[20]   Todd JA, Walker NM, Cooper JD, *et al.* Genetics of Type 1 Diabetes in Finland Robust Wellcome Trust Case Control Consortium associations of four new chromosome regions from genome-wide analyses of type 1 diabetes. Nat Genet 2007; 39(7): 857-64.
[http://dx.doi.org/10.1038/ng2068] [PMID: 17554260]

[21]   Undlien DE, Lie BA, Thorsby E. HLA complex genes in type 1 diabetes and other autoimmune diseases. Which genes are involved? Trends Genet 2001; 17(2): 93-100.
[http://dx.doi.org/10.1016/S0168-9525(00)02180-6] [PMID: 11173119]

[22]   Rich SS. Mapping genes in diabetes. Genetic epidemiological perspective. Diabetes 1990; 39(11): 1315-9.
[http://dx.doi.org/10.2337/diab.39.11.1315] [PMID: 2227105]

[23]   Cooper JD, Howson JMM, Smyth D, *et al.* Type 1 diabetes genetics consortium confirmation of novel type 1 diabetes risk loci in families. Diabetologia 2012; 55(4): 996-1000.
[http://dx.doi.org/10.1007/s00125-012-2450-3] [PMID: 22278338]

[24]   Long SA, Cerosaletti K, Bollyky PL, *et al.* Defects in IL-2R signaling contribute to diminished maintenance of FOXP3 expression in CD4(+)CD25(+) regulatory T-cells of type 1 diabetic subjects. Diabetes 2010; 59(2): 407-15.
[http://dx.doi.org/10.2337/db09-0694] [PMID: 19875613]

[25]   Long SA, Cerosaletti K, Wan JY, *et al.* An autoimmune-associated variant in PTPN2 reveals an impairment of IL-2R signaling in CD4+ T cells. Genes Immun 2011; 12(2): 116-25.
[http://dx.doi.org/10.1038/gene.2010.54] [PMID: 21179116]

[26]   Pugliese A, Zeller M, Fernandez A Jr, *et al.* The insulin gene is transcribed in the human thymus and transcription levels correlate with allelic variation at the INS VNTR-IDDM2 susceptibility locus for type 1 diabetes. Nat Genet 1997; 15(3): 293-7.
[http://dx.doi.org/10.1038/ng0397-293] [PMID: 9054945]

[27]   Sosinowski T, Eisenbarth GS. Type 1 diabetes: primary antigen/peptide/register/trimolecular complex. Immunol Res 2013; 55(1-3): 270-6.

[http://dx.doi.org/10.1007/s12026-012-8367-6] [PMID: 22956469]

[28]   Colli ML, Moore F, Gurzov EN, Ortis F, Eizirik DL. Mda5 and ptpn2, two candidate genes for type 1 diabetes, modify pancreatic β-cell responses to the viral by-product double-stranded rna. Hum Mol Genet 2010; 19(1): 135-46.
[http://dx.doi.org/10.1093/hmg/ddp474] [PMID: 19825843]

[29]   Marroqui L, Dos Santos RS, Fløyel T, *et al.* TYK2, a candidate gene for type 1 diabetes, modulates apoptosis and the innate immune response in human pancreatic β-cells. Diabetes 2015; 64(11): 3808-17.
[http://dx.doi.org/10.2337/db15-0362] [PMID: 26239055]

[30]   Fløyel T, Kaur S, Pociot F. Genes affecting β-cell function in type 1 diabetes. Curr Diab Rep 2015; 15(11): 97.
[http://dx.doi.org/10.1007/s11892-015-0655-9] [PMID: 26391391]

[31]   Santin I, Eizirik DL. Candidate genes for type 1 diabetes modulate pancreatic islet inflammation and β -cell apoptosis. Diabetes Obes Metab 2013; 15(s3) (Suppl. 3): 71-81.
[http://dx.doi.org/10.1111/dom.12162] [PMID: 24003923]

[32]   Törn C, Hadley D, Lee HS, *et al.* Teddy study group role of type 1 diabetes–associated snps on risk of autoantibody positivity in the teddy study. Diabetes 2015; 64(5): 1818-29.
[http://dx.doi.org/10.2337/db14-1497] [PMID: 25422107]

[33]   Steck AK, Wong R, Wagner B, *et al.* Effects of non-hla gene polymorphisms on development of islet autoimmunity and type 1 diabetes in a population with high-risk hla-dr,dq genotypes. Diabetes 2012; 61(3): 753-8.
[http://dx.doi.org/10.2337/db11-1228] [PMID: 22315323]

[34]   Stankov K, Benc D, Draskovic D. Genetic and epigenetic factors in etiology of diabetes mellitus type 1. Pediatrics 2013; 132(6): 1112-22.
[http://dx.doi.org/10.1542/peds.2013-1652] [PMID: 24190679]

[35]   Insel RA, Dunne JL, Atkinson MA, *et al.* Staging presymptomatic type 1 diabetes: A scientific statement of jdrf, the endocrine society, and the american diabetes association. Diabetes Care 2015; 38(10): 1964-74.
[http://dx.doi.org/10.2337/dc15-1419] [PMID: 26404926]

[36]   Shaver KA, Boughman JA, Nance WE. Congenital rubella syndrome and diabetes: A review of epidemiologic, genetic, and immunologic factors. Am Ann Deaf 1985; 130(6): 526-32.
[http://dx.doi.org/10.1353/aad.0.0142] [PMID: 3832941]

[37]   Helminen O, Aspholm S, Pokka T, *et al.* Hba1c predicts time to diagnosis of type 1 diabetes in children at risk. Diabetes 2015; 64(5): 1719-27.
[http://dx.doi.org/10.2337/db14-0497] [PMID: 25524912]

[38]   Norris JM, Barriga K, Klingensmith G, *et al.* Timing of initial cereal exposure in infancy and risk of islet autoimmunity. JAMA 2003; 290(13): 1713-20.
[http://dx.doi.org/10.1001/jama.290.13.1713] [PMID: 14519705]

[39]   Vaarala O, Ilonen J, Ruohtula T, *et al.* Removal of bovine insulin from cow's milk formula and early initiation of beta-cell autoimmunity in the FINDIA pilot study. Arch Pediatr Adolesc Med 2012; 166(7): 608-14.
[http://dx.doi.org/10.1001/archpediatrics.2011.1559] [PMID: 22393174]

[40]   Norris JM, Yin X, Lamb MM, *et al.* Omega-3 polyunsaturated fatty acid intake and islet autoimmunity in children at increased risk for type 1 diabetes. JAMA 2007; 298(12): 1420-8.
[http://dx.doi.org/10.1001/jama.298.12.1420] [PMID: 17895458]

[41]   Willcox A, Richardson SJ, Bone AJ, Foulis AK, Morgan NG. Analysis of islet inflammation in human type 1 diabetes. Clin Exp Immunol 2009; 155(2): 173-81.
[http://dx.doi.org/10.1111/j.1365-2249.2008.03860.x] [PMID: 19128359]

[42] Butler AE, Janson J, Bonner-Weir S, Ritzel R, Rizza RA, Butler PC. Beta-cell deficit and increased beta-cell apoptosis in humans with type 2 diabetes. Diabetes 2003; 52(1): 102-10.
[http://dx.doi.org/10.2337/diabetes.52.1.102] [PMID: 12502499]

[43] Laybutt DR, Kaneto H, Hasenkamp W, *et al.* Increased expression of antioxidant and antiapoptotic genes in islets that may contribute to beta-cell survival during chronic hyperglycemia. Diabetes 2002; 51(2): 413-23.
[http://dx.doi.org/10.2337/diabetes.51.2.413] [PMID: 11812749]

[44] Weir GC, Bonner-Weir S. Five stages of evolving beta-cell dysfunction during progression to diabetes. Diabetes 2004; 53 (Suppl. 3): S16-21.
[http://dx.doi.org/10.2337/diabetes.53.suppl_3.S16] [PMID: 15561905]

[45] Vardi P, Crisa L, Jackson RA, *et al.* Predictive value of intravenous glucose tolerance test insulin secretion less than or greater than the first percentile in islet cell antibody positive relatives of type 1 (insulin-dependent) diabetic patients. Diabetologia 1991; 34(2): 93-102.
[http://dx.doi.org/10.1007/BF00500379] [PMID: 2065854]

[46] Chase HP, Voss MA, Butler-Simon N, Hoops S, O'Brien D, Dobersen MJ. Diagnosis of pre-type I diabetes. J Pediatr 1987; 111(6): 807-12.
[http://dx.doi.org/10.1016/S0022-3476(87)80192-0] [PMID: 3316560]

[47] Srikanta S, Ganda OP, Rabizadeh A, Soeldner JS, Eisenbarth GS. First-degree relatives of patients with type I diabetes mellitus. Islet-cell antibodies and abnormal insulin secretion. N Engl J Med 1985; 313(8): 461-4.
[http://dx.doi.org/10.1056/NEJM198508223130801] [PMID: 3894969]

[48] Ginsberg-Fellner F, Witt ME, Franklin BH, *et al.* Triad of markers for identifying children at high risk of developing insulin-dependent diabetes mellitus. JAMA 1985; 254(11): 1469-72.
[http://dx.doi.org/10.1001/jama.1985.03360110059024] [PMID: 3897593]

[49] Sosenko JM, Skyler JS, Beam CA, *et al.* Type 1 diabetes trialnet and diabetes prevention trial–type 1 study groups acceleration of the loss of the first-phase insulin response during the progression to type 1 diabetes in diabetes prevention trial-type 1 participants. Diabetes 2013; 62(12): 4179-83.
[http://dx.doi.org/10.2337/db13-0656] [PMID: 23863814]

[50] Sosenko JM, Palmer JP, Rafkin LE, *et al.* Diabetes prevention trial-type 1 study group trends of earlier and later responses of c-peptide to oral glucose challenges with progression to type 1 diabetes in diabetes prevention trial-type 1 participants. Diabetes Care 2010; 33(3): 620-5.
[http://dx.doi.org/10.2337/dc09-1770] [PMID: 20032282]

[51] Ferrannini E, Mari A, Nofrate V, Sosenko JM, Skyler JS. Dpt-1 study group progression to diabetes in relatives of type 1 diabetic patients: Mechanisms and mode of onset. Diabetes 2010; 59(3): 679-85.
[http://dx.doi.org/10.2337/db09-1378] [PMID: 20028949]

[52] Sosenko JM, Palmer JP, Rafkin-Mervis L, *et al.* Glucose and C-peptide changes in the perionset period of type 1 diabetes in the Diabetes Prevention Trial-Type 1. Diabetes Care 2008; 31(11): 2188-92.
[http://dx.doi.org/10.2337/dc08-0935] [PMID: 18650369]

[53] Greenbaum CJ, Beam CA, Boulware D, *et al.* Type 1 diabetes trialnet study group fall in c-peptide during first 2 years from diagnosis: Evidence of at least two distinct phases from composite type 1 diabetes trialnet data. Diabetes 2012; 61(8): 2066-73.
[http://dx.doi.org/10.2337/db11-1538] [PMID: 22688329]

[54] Oram RA, Jones AG, Besser REJ, *et al.* The majority of patients with long-duration type 1 diabetes are insulin microsecretors and have functioning beta cells. Diabetologia 2014; 57(1): 187-91.
[http://dx.doi.org/10.1007/s00125-013-3067-x] [PMID: 24121625]

[55] Rosenbloom AL, Hunt SS, Rosenbloom EK, Maclaren NK. Ten-year prognosis of impaired glucose tolerance in siblings of patients with insulin-dependent diabetes. Diabetes 1982; 31(5): 385-7.

[http://dx.doi.org/10.2337/diab.31.5.385] [PMID: 6759254]

[56]   Tarn AC, Smith CP, Spencer KM, Bottazzo GF, Gale EA. Type I (insulin dependent) diabetes: A disease of slow clinical onset? BMJ 1987; 294(6568): 342-5.
[http://dx.doi.org/10.1136/bmj.294.6568.342] [PMID: 3101866]

[57]   Beer SF, Heaton DA, Alberti KGMM, Pyke DA, Leslie RDG. Impaired glucose tolerance precedes but does not predict insulin-dependent diabetes mellitus: A study of identical twins. Diabetologia 1990; 33(8): 497-502.
[http://dx.doi.org/10.1007/BF00405112] [PMID: 2210123]

[58]   Sosenko JM, Skyler JS, Krischer JP, *et al.* Diabetes prevention trial-type 1 study group glucose excursions between states of glycemia with progression to type 1 diabetes in the diabetes prevention trial-type 1 (dpt-1). Diabetes 2010; 59(10): 2386-9.
[http://dx.doi.org/10.2337/db10-0534] [PMID: 20682683]

[59]   Sosenko JM, Palmer JP, Rafkin-Mervis L, *et al.* Diabetes prevention trial-type 1 study group incident dysglycemia and progression to type 1 diabetes among participants in the diabetes prevention trial-type 1. Diabetes Care 2009; 32(9): 1603-7.
[http://dx.doi.org/10.2337/dc08-2140] [PMID: 19487644]

[60]   Sosenko JM, Krischer JP, Palmer JP, *et al.* Diabetes prevention trial-type 1 study group a risk score for type 1 diabetes derived from autoantibody-positive participants in the diabetes prevention trial-type 1. Diabetes Care 2008; 31(3): 528-33.
[http://dx.doi.org/10.2337/dc07-1459] [PMID: 18000175]

[61]   Sosenko JM, Skyler JS, Mahon J, *et al.* Type 1 diabetes trialnet and diabetes prevention trial-type 1 study groups validation of the diabetes prevention trial–type 1 risk score in the trialnet natural history study. Diabetes Care 2011; 34(8): 1785-7.
[http://dx.doi.org/10.2337/dc11-0641] [PMID: 21680724]

[62]   Knip M, Siljander H. Autoimmune mechanisms in type 1 diabetes. Autoimmun Rev 2008; 7(7): 550-7.
[http://dx.doi.org/10.1016/j.autrev.2008.04.008] [PMID: 18625444]

[63]   Kahaly GJ, Hansen MP. Type 1 diabetes associated autoimmunity. Autoimmun Rev 2016; 15(7): 644-8.
[http://dx.doi.org/10.1016/j.autrev.2016.02.017] [PMID: 26903475]

[64]   Atkinson MA, Eisenbarth GS, Michels AW. Type 1 diabetes. Lancet 2014; 383(9911): 69-82.
[http://dx.doi.org/10.1016/S0140-6736(13)60591-7] [PMID: 23890997]

[65]   Bonifacio E. Predicting type 1 diabetes using biomarkers. Diabetes Care 2015; 38(6): 989-96.
[http://dx.doi.org/10.2337/dc15-0101] [PMID: 25998291]

[66]   Vehik K, Lynch KF, Schatz DA, *et al.* Teddy study group reversion of β-cell autoimmunity changes risk of type 1 diabetes: teddy study. Diabetes Care 2016; 39(9): 1535-42.
[http://dx.doi.org/10.2337/dc16-0181] [PMID: 27311490]

[67]   Ozougwu O. The pathogenesis and pathophysiology of type 1 and type 2 diabetes mellitus. J Physiol Pathophysiol 2013; 4(4): 46-57.
[http://dx.doi.org/10.5897/JPAP2013.0001]

[68]   Ziegler AG, Rewers M, Simell O, *et al.* Seroconversion to multiple islet autoantibodies and risk of progression to diabetes in children. JAMA 2013; 309(23): 2473-9.
[http://dx.doi.org/10.1001/jama.2013.6285] [PMID: 23780460]

[69]   Steck AK, Vehik K, Bonifacio E, *et al.* Teddy study group predictors of progression from the appearance of islet autoantibodies to early childhood diabetes: the environmental determinants of diabetes in the young (teddy). Diabetes Care 2015; 38(5): 808-13.
[http://dx.doi.org/10.2337/dc14-2426] [PMID: 25665818]

[70]   Orban T, Sosenko JM, Cuthbertson D, *et al.* Diabetes prevention trial-type 1 study group pancreatic islet autoantibodies as predictors of type 1 diabetes in the diabetes prevention trial-type 1. Diabetes

Care 2009; 32(12): 2269-74.
[http://dx.doi.org/10.2337/dc09-0934] [PMID: 19741189]

[71]   Colman PG, Steele C, Couper JJ, *et al.* Islet autoimmunity in infants with a Type I diabetic relative is common but is frequently restricted to one autoantibody. Diabetologia 2000; 43(2): 203-9.
[http://dx.doi.org/10.1007/s001250050030] [PMID: 10753042]

[72]   Leslie RD, Palmer J, Schloot NC, Lernmark A. Diabetes at the crossroads: Relevance of disease classification to pathophysiology and treatment. Diabetologia 2016; 59(1): 13-20.
[http://dx.doi.org/10.1007/s00125-015-3789-z] [PMID: 26498592]

[73]   Raju SM, Bindu Madala. Illustrated medical biochemistry. New Delhi: Jaypee Brothers Medical Publishers 2010.

[74]   Centers for Disease Control and Prevention National Diabetes Statistics Report website. 2022. Available from : https://www.cdc.gov/diabetes/data/statistics-report/index.html

[75]   Jallut D, Golay A, Munger R, *et al.* Impaired glucose tolerance and diabetes in obesity: A 6-year follow-up study of glucose metabolism. Metabolism 1990; 39(10): 1068-75.
[http://dx.doi.org/10.1016/0026-0495(90)90168-C] [PMID: 2215253]

[76]   Tabák AG, Jokela M, Akbaraly TN, Brunner EJ, Kivimäki M, Witte DR. Trajectories of glycaemia, insulin sensitivity, and insulin secretion before diagnosis of type 2 diabetes: An analysis from the Whitehall II study. Lancet 2009; 373(9682): 2215-21.
[http://dx.doi.org/10.1016/S0140-6736(09)60619-X] [PMID: 19515410]

[77]   DeFronzo RA, Tobin JD, Andres R. Glucose clamp technique: A method for quantifying insulin secretion and resistance. Am J Physiol 1979; 237(3): E214-23.
[PMID: 382871]

[78]   Matthews DR, Hosker JP, Rudenski AS, Naylor BA, Treacher DF, Turner RC. Homeostasis model assessment: insulin resistance and? -cell function from fasting plasma glucose and insulin concentrations in man. Diabetologia 1985; 28(7): 412-9.
[http://dx.doi.org/10.1007/BF00280883] [PMID: 3899825]

[79]   Himsworth H, Kerr RB. Insulin-sensitive and insulin-insensitive types of diabetes mellitus. Clin Sci 1939; 4: 119-52.

[80]   Warram JH, Martin BC, Krolewski AS, Soeldner JS, Kahn CR. Slow glucose removal rate and hyperinsulinemia precede the development of type II diabetes in the offspring of diabetic parents. Ann Intern Med 1990; 113(12): 909-15.
[http://dx.doi.org/10.7326/0003-4819-113-12-909] [PMID: 2240915]

[81]   Lillioja S, Mott DM, Howard BV, *et al.* Impaired glucose tolerance as a disorder of insulin action. Longitudinal and cross-sectional studies in Pima Indians. N Engl J Med 1988; 318(19): 1217-25.
[http://dx.doi.org/10.1056/NEJM198805123181901] [PMID: 3283552]

[82]   Haffner SM, Stern MP, Dunn J, Mobley M, Blackwell J, Bergman RN. Diminished insulin sensitivity and increased insulin response in nonobese, nondiabetic Mexican Americans. Metabolism 1990; 39(8): 842-7.
[http://dx.doi.org/10.1016/0026-0495(90)90130-5] [PMID: 2198435]

[83]   Okada T, Liew CW, Hu J, *et al.* Insulin receptors in β-cells are critical for islet compensatory growth response to insulin resistance. Proc Natl Acad Sci USA 2007; 104(21): 8977-82.
[http://dx.doi.org/10.1073/pnas.0608703104] [PMID: 17416680]

[84]   Kulkarni RN, Brüning JC, Winnay JN, Postic C, Magnuson MA, Kahn CR. Tissue-specific knockout of the insulin receptor in pancreatic β cells creates an insulin secretory defect similar to that in type 2 diabetes. Cell 1999; 96(3): 329-39.
[http://dx.doi.org/10.1016/S0092-8674(00)80546-2] [PMID: 10025399]

[85]   Kobayashi M, Kikuchi O, Sasaki T, *et al.* FoxO1 as a double-edged sword in the pancreas: analysis of pancreas- and β-cell-specific FoxO1 knockout mice. Am J Physiol Endocrinol Metab 2012; 302(5):

E603-13.
[http://dx.doi.org/10.1152/ajpendo.00469.2011] [PMID: 22215655]

[86]   Talchai C, Xuan S, Lin HV, Sussel L, Accili D. Pancreatic β cell dedifferentiation as a mechanism of diabetic β cell failure. Cell 2012; 150(6): 1223-34.
[http://dx.doi.org/10.1016/j.cell.2012.07.029] [PMID: 22980982]

[87]   Cinti F, Bouchi R, Kim-Muller JY, *et al.* Evidence of β-cell dedifferentiation in human type 2 diabetes. J Clin Endocrinol Metab 2016; 101(3): 1044-54.
[http://dx.doi.org/10.1210/jc.2015-2860] [PMID: 26713822]

[88]   Kawamori D, Kurpad AJ, Hu J, *et al.* Insulin signaling in alpha cells modulates glucagon secretion *in vivo*. Cell Metab 2009; 9(4): 350-61.
[http://dx.doi.org/10.1016/j.cmet.2009.02.007] [PMID: 19356716]

[89]   Laakso M, Edelman SV, Brechtel G, Baron AD. Impaired insulin-mediated skeletal muscle blood flow in patients with niddm. Diabetes 1992; 41(9): 1076-83.
[http://dx.doi.org/10.2337/diab.41.9.1076] [PMID: 1499861]

[90]   Laakso M, Edelman SV, Brechtel G, Baron AD. Decreased effect of insulin to stimulate skeletal muscle blood flow in obese man. A novel mechanism for insulin resistance. J Clin Invest 1990; 85(6): 1844-52.
[http://dx.doi.org/10.1172/JCI114644] [PMID: 2189893]

[91]   Clark MG, Wallis MG, Barrett EJ, *et al.* Blood flow and muscle metabolism: A focus on insulin action. Am J Physiol Endocrinol Metab 2003; 284(2): E241-58.
[http://dx.doi.org/10.1152/ajpendo.00408.2002] [PMID: 12531739]

[92]   Konishi M, Sakaguchi M, Lockhart SM, *et al.* Endothelial insulin receptors differentially control insulin signaling kinetics in peripheral tissues and brain of mice. Proc Natl Acad Sci USA 2017; 114(40): E8478-87.
[http://dx.doi.org/10.1073/pnas.1710625114] [PMID: 28923931]

[93]   Rask-Madsen C, Li Q, Freund B, *et al.* Loss of insulin signaling in vascular endothelial cells accelerates atherosclerosis in apolipoprotein E null mice. Cell Metab 2010; 11(5): 379-89.
[http://dx.doi.org/10.1016/j.cmet.2010.03.013] [PMID: 20444418]

[94]   Tsuchiya K, Tanaka J, Shuiqing Y, *et al.* FoxOs integrate pleiotropic actions of insulin in vascular endothelium to protect mice from atherosclerosis. Cell Metab 2012; 15(3): 372-81.
[http://dx.doi.org/10.1016/j.cmet.2012.01.018] [PMID: 22405072]

[95]   Park K, Mima A, Li Q, *et al.* Insulin decreases atherosclerosis by inducing endothelin receptor B expression. JCI Insight 2016; 1(6)e86574.
[http://dx.doi.org/10.1172/jci.insight.86574] [PMID: 27200419]

[96]   O'Neill BT, Kim J, Wende AR, *et al.* A conserved role for phosphatidylinositol 3-kinase but not Akt signaling in mitochondrial adaptations that accompany physiological cardiac hypertrophy. Cell Metab 2007; 6(4): 294-306.
[http://dx.doi.org/10.1016/j.cmet.2007.09.001] [PMID: 17908558]

[97]   Belke DD, Betuing S, Tuttle MJ, *et al.* Insulin signaling coordinately regulates cardiac size, metabolism, and contractile protein isoform expression. J Clin Invest 2002; 109(5): 629-39.
[http://dx.doi.org/10.1172/JCI0213946] [PMID: 11877471]

[98]   Mazumder PK, O'Neill BT, Roberts MW, *et al.* Impaired cardiac efficiency and increased fatty acid oxidation in insulin-resistant ob/ob mouse hearts. Diabetes 2004; 53(9): 2366-74.
[http://dx.doi.org/10.2337/diabetes.53.9.2366] [PMID: 15331547]

[99]   Moore KJ, Tabas I. Macrophages in the pathogenesis of atherosclerosis. Cell 2011; 145(3): 341-55.
[http://dx.doi.org/10.1016/j.cell.2011.04.005] [PMID: 21529710]

[100]   Badiu C. Williams Textbook of Endocrinology. Bucharest: Acta Endocrinologica 2019; p. 1359.

[101] Cefalu WT, Werbel S, Bell-Farrow AD, *et al.* Insulin resistance and fat patterning with aging: Relationship to metabolic risk factors for cardiovascular disease. Metabolism 1998; 47(4): 401-8. [http://dx.doi.org/10.1016/S0026-0495(98)90050-6] [PMID: 9550536]

[102] Larsson B, Svärdsudd K, Welin L, Wilhelmsen L, Björntorp P, Tibblin G. Abdominal adipose tissue distribution, obesity, and risk of cardiovascular disease and death: 13 year follow up of participants in the study of men born in 1913. BMJ 1984; 288(6428): 1401-4. [http://dx.doi.org/10.1136/bmj.288.6428.1401] [PMID: 6426576]

[103] Després JP, Tremblay A, Pérusse L, Leblanc C, Bouchard C. Abdominal adipose tissue and serum HDL-cholesterol: association independent from obesity and serum triglyceride concentration. Int J Obes 1988; 12(1): 1-13. [PMID: 2966132]

[104] Landin K, Krotkiewski M, Smith U. Importance of obesity for the metabolic abnormalities associated with an abdominal fat distribution. Metabolism 1989; 38(6): 572-6. [http://dx.doi.org/10.1016/0026-0495(89)90219-9] [PMID: 2657328]

[105] Heitmann BL. The variation in blood lipid levels described by various measures of overall and abdominal obesity in Danish men and women aged 35-65 years. Eur J Clin Nutr 1992; 46(8): 597-605. [PMID: 1396477]

[106] Reeder BA, Senthilselvan A, Després JP, *et al.* Canadian heart health surveys research group the association of cardiovascular disease risk factors with abdominal obesity in canada. CMAJ 1997; 157 (Suppl. 1): S39-45. [PMID: 9220953]

[107] Lamarche B. Abdominal obesity and its metabolic complications. Coron Artery Dis 1998; 9(8): 473-82. [http://dx.doi.org/10.1097/00019501-199809080-00002] [PMID: 9847978]

[108] McLaughlin T, Lamendola C, Liu A, Abbasi F. Preferential fat deposition in subcutaneous versus visceral depots is associated with insulin sensitivity. J Clin Endocrinol Metab 2011; 96(11): E1756-60. [http://dx.doi.org/10.1210/jc.2011-0615] [PMID: 21865361]

[109] Arner P, Hellström L, Wahrenberg H, Brönnegård M. Beta-adrenoceptor expression in human fat cells from different regions. J Clin Invest 1990; 86(5): 1595-600. [http://dx.doi.org/10.1172/JCI114880] [PMID: 2173724]

[110] Mittelman SD, Van Citters GW, Kim SP, *et al.* Longitudinal compensation for fat-induced insulin resistance includes reduced insulin clearance and enhanced beta-cell response. Diabetes 2000; 49(12): 2116-25. [http://dx.doi.org/10.2337/diabetes.49.12.2116] [PMID: 11118015]

[111] DeFronzo RA. Pathogenesis of type 2 diabetes mellitus. Med Clin North Am 2004; 88(4): 787-835. [http://dx.doi.org/10.1016/j.mcna.2004.04.013] [PMID: 15308380]

[112] Pilkis SJ, Granner DK. Molecular physiology of the regulation of hepatic gluconeogenesis and glycolysis. Annu Rev Physiol 1992; 54(1): 885-909. [http://dx.doi.org/10.1146/annurev.ph.54.030192.004321] [PMID: 1562196]

[113] DeFronzo RA. Lilly lecture 1987. The triumvirate: β-cell, muscle, liver. A collusion responsible for NIDDM. Diabetes 1988; 37(6): 667-87. [http://dx.doi.org/10.2337/diab.37.6.667] [PMID: 3289989]

[114] DeFronzo RA, Ferrannini E, Simonson DC. Fasting hyperglycemia in non-insulin-dependent diabetes mellitus: Contributions of excessive hepatic glucose production and impaired tissue glucose uptake. Metabolism 1989; 38(4): 387-95. [http://dx.doi.org/10.1016/0026-0495(89)90129-7] [PMID: 2657323]

[115] Birkenfeld AL, Shulman GI. Nonalcoholic fatty liver disease, hepatic insulin resistance, and type 2 Diabetes. Hepatology 2014; 59(2): 713-23.

[http://dx.doi.org/10.1002/hep.26672] [PMID: 23929732]

[116]   Taylor R. Pathogenesis of type 2 diabetes: tracing the reverse route from cure to cause. Diabetologia 2008; 51(10): 1781-9.
[http://dx.doi.org/10.1007/s00125-008-1116-7] [PMID: 18726585]

[117]   Sattar N, Gill JMR. Type 2 diabetes as a disease of ectopic fat? BMC Med 2014; 12(1): 123.
[http://dx.doi.org/10.1186/s12916-014-0123-4] [PMID: 25159817]

[118]   Defronzo R, Deibert D, Hendler R, Felig P, Soman V. Insulin sensitivity and insulin binding to monocytes in maturity-onset diabetes. J Clin Invest 1979; 63(5): 939-46.
[http://dx.doi.org/10.1172/JCI109394] [PMID: 376552]

[119]   Acierno C, Caturano A, Pafundi PC, Nevola R, Adinolfi LE, Sasso FC. Nonalcoholic fatty liver disease and type 2 diabetes: Pathophysiological mechanisms shared between the two faces of the same coin. Explor Med 2020; 1: 287-306.
[http://dx.doi.org/10.37349/emed.2020.00019]

[120]   Ratziu V, Goodman Z, Sanyal A. Current efforts and trends in the treatment of NASH. J Hepatol 2015; 62(1) (Suppl.): S65-75.
[http://dx.doi.org/10.1016/j.jhep.2015.02.041] [PMID: 25920092]

[121]   Williams CD, Stengel J, Asike MI, *et al.* Prevalence of nonalcoholic fatty liver disease and nonalcoholic steatohepatitis among a largely middle-aged population utilizing ultrasound and liver biopsy: a prospective study. Gastroenterology 2011; 140(1): 124-31.
[http://dx.doi.org/10.1053/j.gastro.2010.09.038] [PMID: 20858492]

[122]   Loomba R, Abraham M, Unalp A, *et al.* Nonalcoholic steatohepatitis clinical research network association between diabetes, family history of diabetes, and risk of nonalcoholic steatohepatitis and fibrosis. Hepatology 2012; 56(3): 943-51.
[http://dx.doi.org/10.1002/hep.25772] [PMID: 22505194]

[123]   Mantovani A, Byrne CD, Bonora E, Targher G. Nonalcoholic fatty liver disease and risk of incident type 2 diabetes: A meta-analysis. Diabetes Care 2018; 41(2): 372-82.
[http://dx.doi.org/10.2337/dc17-1902] [PMID: 29358469]

[124]   Marchesini G, Brizi M, Morselli-Labate AM, *et al.* Association of nonalcoholic fatty liver disease with insulin resistance. Am J Med 1999; 107(5): 450-5.
[http://dx.doi.org/10.1016/S0002-9343(99)00271-5] [PMID: 10569299]

[125]   Trombetta M, Spiazzi G, Zoppini G, Muggeo M. Review article: Type 2 diabetes and chronic liver disease in the Verona diabetes study. Aliment Pharmacol Ther 2005; 22(s2) (Suppl. 2): 24-7.
[http://dx.doi.org/10.1111/j.1365-2036.2005.02590.x] [PMID: 16225467]

[126]   de Marco R, Locatelli F, Zoppini G, Verlato G, Bonora E, Muggeo M. Cause-specific mortality in type 2 diabetes. The Verona Diabetes Study. Diabetes Care 1999; 22(5): 756-61.
[http://dx.doi.org/10.2337/diacare.22.5.756] [PMID: 10332677]

[127]   Adinolfi LE, Petta S, Fracanzani AL, *et al.* Impact of hepatitis C virus clearance by direct-acting antiviral treatment on the incidence of major cardiovascular events: A prospective multicentre study. Atherosclerosis 2020; 296: 40-7.
[http://dx.doi.org/10.1016/j.atherosclerosis.2020.01.010] [PMID: 32005004]

[128]   Minutolo R, Ferdinando Carlo Sasso, Chiodini P, Cianciaruso B, Carbonara O, Zamboli P. Management of cardiovascular risk factors in advanced type 2 diabetic nephropathy : A comparative analysis in nephrology, diabetology and primary care settings. J Hypertens 2006; 24(8): 1655-61.

[129]   Samuel VT, Shulman GI. The pathogenesis of insulin resistance: Integrating signaling pathways and substrate flux. J Clin Invest 2016; 126(1): 12-22.
[http://dx.doi.org/10.1172/JCI77812] [PMID: 26727229]

[130]   Vidal-Puig A. Adipose tissue expandability, lipotoxicity and the metabolic syndrome. Endocrinol Nutr 2013; 60 (Suppl. 1): 39-43.

[http://dx.doi.org/10.1016/S1575-0922(13)70026-3] [PMID: 24490226]

[131] Iyer A, Fairlie DP, Prins JB, Hammock BD, Brown L. Inflammatory lipid mediators in adipocyte function and obesity. Nat Rev Endocrinol 2010; 6(2): 71-82.
       [http://dx.doi.org/10.1038/nrendo.2009.264] [PMID: 20098448]

[132] Karastergiou K, Mohamed-Ali V. The autocrine and paracrine roles of adipokines. Mol Cell Endocrinol 2010; 318(1-2): 69-78.
       [http://dx.doi.org/10.1016/j.mce.2009.11.011] [PMID: 19948207]

[133] Shreiner AB, Kao JY, Young VB. The gut microbiome in health and in disease. Curr Opin Gastroenterol 2015; 31(1): 69-75.
       [http://dx.doi.org/10.1097/MOG.0000000000000139] [PMID: 25394236]

[134] Ussar S, Griffin NW, Bezy O, *et al.* Interactions between gut microbiota, host genetics and diet modulate the predisposition to obesity and metabolic syndrome. Cell Metab 2015; 22(3): 516-30.
       [http://dx.doi.org/10.1016/j.cmet.2015.07.007] [PMID: 26299453]

[135] Lynch SV, Pedersen O. The human intestinal microbiome in health and disease. N Engl J Med 2016; 375(24): 2369-79.
       [http://dx.doi.org/10.1056/NEJMra1600266]

[136] Blaser MJ. The microbiome revolution. J Clin Invest 2014; 124(10): 4162-5.
       [http://dx.doi.org/10.1172/JCI78366] [PMID: 25271724]

[137] Rooks MG, Garrett WS. Gut microbiota, metabolites and host immunity. Nat Rev Immunol 2016; 16(6): 341-52.
       [http://dx.doi.org/10.1038/nri.2016.42] [PMID: 27231050]

[138] Schirmer M, Smeekens SP, Vlamakis H, *et al.* Linking the human gut microbiome to inflammatory cytokine production capacity. Cell 2016; 167(4): 1125-1136.e8.
       [http://dx.doi.org/10.1016/j.cell.2016.10.020] [PMID: 27814509]

[139] Burcelin R, Serino M, Chabo C, Blasco-Baque V, Amar J. Gut microbiota and diabetes: From pathogenesis to therapeutic perspective. Acta Diabetol 2011; 48(4): 257-73.
       [http://dx.doi.org/10.1007/s00592-011-0333-6] [PMID: 21964884]

[140] Pedersen HK, Gudmundsdottir V, Nielsen HB, *et al.* Metahit consortium human gut microbes impact host serum metabolome and insulin sensitivity. Nature 2016; 535(7612): 376-81.
       [http://dx.doi.org/10.1038/nature18646] [PMID: 27409811]

[141] Schroeder BO, Bäckhed F. Signals from the gut microbiota to distant organs in physiology and disease. Nat Med 2016; 22(10): 1079-89.
       [http://dx.doi.org/10.1038/nm.4185] [PMID: 27711063]

[142] Walford GA, Davis J, Warner AS, *et al.* Branched chain and aromatic amino acids change acutely following two medical therapies for type 2 diabetes mellitus. Metabolism 2013; 62(12): 1772-8.
       [http://dx.doi.org/10.1016/j.metabol.2013.07.003] [PMID: 23953891]

[143] Drucker DJ. The biology of incretin hormones. Cell Metab 2006; 3(3): 153-65.
       [http://dx.doi.org/10.1016/j.cmet.2006.01.004] [PMID: 16517403]

[144] Rask E, Olsson T, Söderberg S, *et al.* Impaired incretin response after a mixed meal is associated with insulin resistance in nondiabetic men. Diabetes Care 2001; 24(9): 1640-5.
       [http://dx.doi.org/10.2337/diacare.24.9.1640] [PMID: 11522713]

[145] Muscelli E, Mari A, Casolaro A, *et al.* Separate impact of obesity and glucose tolerance on the incretin effect in normal subjects and type 2 diabetic patients. Diabetes 2008; 57(5): 1340-8.
       [http://dx.doi.org/10.2337/db07-1315] [PMID: 18162504]

[146] Højberg PV, Vilsbøll T, Rabøl R, *et al.* Four weeks of near-normalisation of blood glucose improves the insulin response to glucagon-like peptide-1 and glucose-dependent insulinotropic polypeptide in patients with type 2 diabetes. Diabetologia 2009; 52(2): 199-207.

[http://dx.doi.org/10.1007/s00125-008-1195-5] [PMID: 19037628]

[147]   MacLean H. Some observations on diabetes and insulin in general practice. Postgrad Med J 1926; 1(6): 73-7.
[http://dx.doi.org/10.1136/pgmj.1.6.73] [PMID: 21312442]

[148]   Abdul-Ghani MA, DeFronzo RA. Inhibition of renal glucose reabsorption: A novel strategy for achieving glucose control in type 2 diabetes mellitus. Endocr Pract 2008; 14(6): 782-90.
[http://dx.doi.org/10.4158/EP.14.6.782] [PMID: 18996802]

[149]   Hasan FM, Alsahli M, Gerich JE. SGLT2 inhibitors in the treatment of type 2 diabetes. Diabetes Res Clin Pract 2014; 104(3): 297-322.
[http://dx.doi.org/10.1016/j.diabres.2014.02.014] [PMID: 24735709]

[150]   Farber SJ, Berger EY, Earle DP. Effect of diabetes and insulin of the maximum capacity of the renal tubules to reabsorb glucose. J Clin Invest 1951; 30(2): 125-9.
[http://dx.doi.org/10.1172/JCI102424] [PMID: 14814204]

[151]   Rahmoune H, Thompson PW, Ward JM, Smith CD, Hong G, Brown J. Glucose transporters in human renal proximal tubular cells isolated from the urine of patients with non-insulin-dependent diabetes. Diabetes 2005; 54(12): 3427-34.
[http://dx.doi.org/10.2337/diabetes.54.12.3427] [PMID: 16306358]

[152]   Saad MF, Knowler WC, Pettitt DJ, Nelson RG, Mott DM, Bennett PH. The natural history of impaired glucose tolerance in the Pima Indians. N Engl J Med 1988; 319(23): 1500-6.
[http://dx.doi.org/10.1056/NEJM198812083192302] [PMID: 3054559]

[153]   Lillioja S, Mott DM, Spraul M, *et al.* Insulin resistance and insulin secretory dysfunction as precursors of non-insulin-dependent diabetes mellitus. Prospective studies of Pima Indians. N Engl J Med 1993; 329(27): 1988-92.
[http://dx.doi.org/10.1056/NEJM199312303292703] [PMID: 8247074]

[154]   Haffner SM, Miettinen H, Gaskill SP, Stern MP. Decreased insulin secretion and increased insulin resistance are independently related to the 7-year risk of NIDDM in Mexican-Americans. Diabetes 1995; 44(12): 1386-91.
[http://dx.doi.org/10.2337/diab.44.12.1386] [PMID: 7589843]

[155]   Sicree RA, Zimmet PZ, King HOM, Coventry JS. Plasma insulin response among Nauruans. Prediction of deterioration in glucose tolerance over 6 yr. Diabetes 1987; 36(2): 179-86.
[http://dx.doi.org/10.2337/diab.36.2.179] [PMID: 3542644]

[156]   Xu E, Kumar M, Zhang Y, *et al.* Intra-islet insulin suppresses glucagon release via GABA-GABAA receptor system. Cell Metab 2006; 3(1): 47-58.
[http://dx.doi.org/10.1016/j.cmet.2005.11.015] [PMID: 16399504]

[157]   Frayling TM. Genome–wide association studies provide new insights into type 2 diabetes aetiology. Nat Rev Genet 2007; 8(9): 657-62.
[http://dx.doi.org/10.1038/nrg2178] [PMID: 17703236]

[158]   Zeggini E, Scott LJ, Saxena R, *et al.* Wellcome trust case control consortium meta-analysis of genome-wide association data and large-scale replication identifies additional susceptibility loci for type 2 diabetes. Nat Genet 2008; 40(5): 638-45.
[http://dx.doi.org/10.1038/ng.120] [PMID: 18372903]

[159]   Tiberti C, Giordano C, Locatelli M, *et al.* Identification of tyrosine phosphatase 2(256-760) construct as a new, sensitive marker for the detection of islet autoimmunity in type 2 diabetic patients: The non-insulin requiring autoimmune diabetes (NIRAD) study 2. Diabetes 2008; 57(5): 1276-83.
[http://dx.doi.org/10.2337/db07-0874] [PMID: 18332100]

[160]   Brooks-Worrell BM, Boyko EJ, Palmer JP. Impact of islet autoimmunity on the progressive β-cell functional decline in type 2 diabetes. Diabetes Care 2014; 37(12): 3286-93.
[http://dx.doi.org/10.2337/dc14-0961] [PMID: 25239783]

[161] Chan JM, Rimm EB, Colditz GA, Stampfer MJ, Willett WC. Obesity, fat distribution, and weight gain as risk factors for clinical diabetes in men. Diabetes Care 1994; 17(9): 961-9.
[http://dx.doi.org/10.2337/diacare.17.9.961] [PMID: 7988316]

[162] Colditz GA, Willett WC, Rotnitzky A, Manson JE. Weight gain as a risk factor for clinical diabetes mellitus in women. Ann Intern Med 1995; 122(7): 481-6.
[http://dx.doi.org/10.7326/0003-4819-122-7-199504010-00001] [PMID: 7872581]

[163] van der Zijl NJ, Goossens GH, Moors CCM, *et al.* Ectopic fat storage in the pancreas, liver, and abdominal fat depots: impact on β-cell function in individuals with impaired glucose metabolism. J Clin Endocrinol Metab 2011; 96(2): 459-67.
[http://dx.doi.org/10.1210/jc.2010-1722] [PMID: 21084401]

[164] Araneta MRG, Kanaya AM, Hsu WC, *et al.* Optimum BMI cut points to screen asian americans for type 2 diabetes. Diabetes Care 2015; 38(5): 814-20.
[http://dx.doi.org/10.2337/dc14-2071] [PMID: 25665815]

[165] Henry RR, Wallace P, Olefsky JM. Effects of weight loss on mechanisms of hyperglycemia in obese non-insulin-dependent diabetes mellitus. Diabetes 1986; 35(9): 990-8.
[http://dx.doi.org/10.2337/diab.35.9.990] [PMID: 3527829]

[166] Lim EL, Hollingsworth KG, Aribisala BS, Chen MJ, Mathers JC, Taylor R. Reversal of type 2 diabetes: normalisation of beta cell function in association with decreased pancreas and liver triacylglycerol. Diabetologia 2011; 54(10): 2506-14.
[http://dx.doi.org/10.1007/s00125-011-2204-7] [PMID: 21656330]

[167] McCaffery JM, Jablonski KA, Franks PW, Dagogo-Jack S, Wing RR, Knowler WC, *et al.* TCF7L2 polymorphism, weight loss and proinsulin Insulin ratio in the diabetes prevention program. PLoS One. 2011; 6: p. (7)e21518.

[168] Knutson KL, Van Cauter E. Associations between sleep loss and increased risk of obesity and diabetes. Ann N Y Acad Sci 2008; 1129(1): 287-304.
[http://dx.doi.org/10.1196/annals.1417.033] [PMID: 18591489]

[169] Maury E, Ramsey KM, Bass J. Circadian rhythms and metabolic syndrome: From experimental genetics to human disease. Circ Res 2010; 106(3): 447-62.
[http://dx.doi.org/10.1161/CIRCRESAHA.109.208355] [PMID: 20167942]

[170] Kaur A, Mokhlesi B. The Effect of OSA Therapy on Glucose Metabolism: It's All about CPAP Adherence! J Clin Sleep Med 2017; 13(3): 365-7.
[http://dx.doi.org/10.5664/jcsm.6480] [PMID: 28212697]

[171] Basu A, Dalla Man C, Basu R, Toffolo G, Cobelli C, Rizza RA. Effects of type 2 diabetes on insulin secretion, insulin action, glucose effectiveness, and postprandial glucose metabolism. Diabetes Care 2009; 32(5): 866-72.
[http://dx.doi.org/10.2337/dc08-1826] [PMID: 19196896]

[172] van Haeften TW, Pimenta W, Mitrakou A, *et al.* Relative contributions of β-cell function and tissue insulin sensitivity to fasting and postglucose-load glycemia. Metabolism 2000; 49(10): 1318-25.
[http://dx.doi.org/10.1053/meta.2000.9526] [PMID: 11079822]

[173] Yoshioka N, Kuzuya T, Matsuda A, Taniguchi M, Iwamoto Y. Serum proinsulin levels at fasting and after oral glucose load in patients with Type 2 (non-insulin-dependent) diabetes mellitus. Diabetologia 1988; 31(6): 355-60.
[http://dx.doi.org/10.1007/BF02341503] [PMID: 3046976]

[174] Giugliano D, Maiorino MI, Bellastella G, Esposito K. Comment on American Diabetes Association. Approaches to Glycemic Treatment. Sec. 7. In *Standards of Medical Care in Diabetes—2016.* Diabetes Care 2016;39(Suppl. 1):S52–S59. Diabetes Care 2016; 39(6): e86-7.
[http://dx.doi.org/10.2337/dc15-2829] [PMID: 27222559]

[175] Phillips LS, Ratner RE, Buse JB, Kahn SE. We can change the natural history of type 2 diabetes.

Diabetes Care 2014; 37(10): 2668-76.
[http://dx.doi.org/10.2337/dc14-0817] [PMID: 25249668]

[176]   Bergman M, Dankner R, Roth J, Narayan KMV. Are current diagnostic guidelines delaying early detection of dysglycemic states? Time for new approaches. Endocrine 2013; 44(1): 66-9.
[http://dx.doi.org/10.1007/s12020-013-9873-6] [PMID: 23325362]

[177]   Phillips LS, Twombly JG. It's time to overcome clinical inertia. Ann Intern Med 2008; 148(10): 783-5.
[http://dx.doi.org/10.7326/0003-4819-148-10-200805200-00011] [PMID: 18490691]

[178]   Nichols GA, Koo YH, Shah SN. Delay of insulin addition to oral combination therapy despite inadequate glycemic control: delay of insulin therapy. J Gen Intern Med 2007; 22(4): 453-8.
[http://dx.doi.org/10.1007/s11606-007-0139-y] [PMID: 17372792]

[179]   Khunti K, Wolden ML, Thorsted BL, Andersen M, Davies MJ. Clinical inertia in people with type 2 diabetes: a retrospective cohort study of more than 80,000 people. Diabetes Care 2013; 36(11): 3411-7.
[http://dx.doi.org/10.2337/dc13-0331] [PMID: 23877982]

[180]   Available from: https://digital.nhs.uk/data-and-information/publications/statistical/national-diabe-es-audit/national-diabetes-audit-2011-12

[181]   Fajans SS, Bell GI. MODY: history, genetics, pathophysiology, and clinical decision making. Diabetes Care 2011; 34(8): 1878-84.
[http://dx.doi.org/10.2337/dc11-0035] [PMID: 21788644]

[182]   Thanabalasingham G, Owen KR. Diagnosis and management of maturity onset diabetes of the young (MODY). BMJ. 2011; 19: 343.

[183]   Kobayashi T, Tamemoto K, Nakanishi K, *et al.* Immunogenetic and clinical characterization of slowly progressive IDDM. Diabetes Care 1993; 16(5): 780-8.
[http://dx.doi.org/10.2337/diacare.16.5.780] [PMID: 8098691]

[184]   Leslie RDG, Williams R, Pozzilli P. Clinical review: Type 1 diabetes and latent autoimmune diabetes in adults: one end of the rainbow. J Clin Endocrinol Metab 2006; 91(5): 1654-9.
[http://dx.doi.org/10.1210/jc.2005-1623] [PMID: 16478821]

[185]   Naik RG, Palmer JP. Latent autoimmune diabetes in adults (LADA). Rev Endocr Metab Disord 2003; 4(3): 233-41.
[http://dx.doi.org/10.1023/A:1025148211587] [PMID: 14501174]

[186]   Pozzilli P, Guglielmi C, Caprio S, Buzzetti R. Obesity, autoimmunity, and double diabetes in youth. Diabetes Care 2011; 34(Suppl 2) (Suppl. 2): S166-70.
[http://dx.doi.org/10.2337/dc11-s213] [PMID: 21525450]

CHAPTER 2

Treatment Approaches and Challenges

Ramin Malboosbaf[1,*] and **Neda Hatami**[1]

[1] *Endocrine Research Center, Institute of Endocrinology and Metabolism, Iran University of Medical Sciences, Tehran, Iran*

Abstract: Diabetes drugs are given in monotherapy or in combination. The significant challenges in effective diabetes management are optimizing current treatments to ensure optimal and stable glucose control with minimal side effects and reducing long-term complications of diabetes. This chapter reviews these conventional drugs with their mechanism of action, side effects, and efficacy and safety profile.

Keywords: Diabetes, Disease, Safety, Treatment.

INTRODUCTION

Many people worldwide are affected by diabetes mellitus (DM), a significant public health problem [1]. The worldwide increase in diabetes patients, maybe primarily attributable to the trend toward sedentary living [2]. Retinopathy, nephropathy, neuropathy, and cardiovascular complications are DM-related complications [3].

Diabetes drugs are given in monotherapy or combination [4]. The significant challenges in effective diabetes management are optimizing current treatments to ensure optimal and stable glucose control with minimal side effects and reducing long-term complications of diabetes [5]. Nanoformulations can solve some of the disadvantages of current anti-diabetic drugs [5] and, more importantly, promote cellular uptake and enhance the pharmacokinetics and pharmacodynamics of drugs [6].

---

* **Corresponding author Ramin Malboosbaf:** Endocrine Research Center, Institute of Endocrinology and Metabolism, Iran University of Medical Sciences, Tehran, Iran; E-mail: malboosbaf.r@gmail.com

PHARMACOLOGIC THERAPY FOR ADULTS WITH TYPE 1 DIABETES

Insulin Therapy

Insulin treatment is essential for these individuals because the absence or near-absence of β-cell function is the hallmark of T1DM [7]. Furthermore, during the past three decades, there has been growing evidence that the optimal combination of effectiveness and safety for patients with T1DM is provided by more intense insulin replacement, such as numerous daily insulin injections or continuous subcutaneous delivery *via* an insulin pump [8 - 10].

Basal insulin, prandial insulin, and correction insulin are frequently used in insulin replacement therapy [11]. NPH insulin, long-acting insulin analogs, and continuous rapid-acting insulin delivery *via* an insulin pump are all components of basal insulin. Compared to NPH insulin, basal insulin analogs have a longer duration of action and plasma concentration and activity profiles that are flatter and more constant. Compared to standard human insulin, rapid-acting analogs (RAA) have a quicker onset, peak, and duration of action. Compared to human insulin, treatment with insulin analogs is associated with lower HbA1C and less hypoglycemia, weight gain, and hypoglycemia in T1DM patients [12 - 14]. Compared to RAA, inhaled human insulin may cause less hypoglycemia and weight gain due to its rapid peak and shorter action duration [15]. Recently, two new formulations of injectable insulin were released with improved fast-acting profiles. Faster-acting insulin aspart and insulin lispro-aabc are better at reducing prandial excursions than RAA [16, 17]. In addition, compared to U-100 glargine, longer-acting basal analogs (U-300 glargine or degludec) may reduce the risk of hypoglycemia in T1DM patients [18, 19]. Despite the advantages of insulin analogs for T1DM patients, some people cannot afford the expense and level of care needed to utilize them [20].

There are numerous insulin treatment options. In order to prevent diabetic ketoacidosis, avoid severe hypoglycemia, and meet individual glycemic goals, the administration of some form of insulin in a planned, individualized regimen is essential for T1DM treatment [20]. By altering the original insulin molecule and changing its constituent parts, several other forms of insulin molecules have developed [20]. These insulin analogs' pharmacodynamic and pharmacokinetic profiles are diverse (Table **1**). Based on their pharmacokinetic and pharmacodynamic properties, these insulins are characterized and administered [21].

Table 1. Different types of insulins.

Insulin Type	Examples	Onset of Action (min)	Time to Peak (hours)	Duration (hours)	Administration
Rapid-acting	Aspart Lispro Glulisine	10-20	0.5-1.5	3-5	0-15min before or just after meals
Short-acting	Regular human	30-45	2-4	4-8	15-30min before meals
Intermediate-acting	NPH	60-120	4-8	12-20	Once or twice daily
Long-acting	Detemir	60-120	6-10	16-24	Usually once daily
	Glargine	60-120	No pronounced peak	~24	
	Degludec	60-120	No pronounced peak	Up to 72	
premixed	70/30NPH/R	30-40	4-8	12-20	Usually twice daily, 0-30min before meals
	70/30 protamine-aspart/aspart	10-20			
Concentrated	U-300 glargine	60-120	No pronounced peak	Up to 72	Once daily
	U-500 human regular	30-45	6-12	12-24	Twice daily
	U-200 degludec	60-120	No pronounced peak	>24	Once daily

There are numerous drawbacks to conventional prandial and basal insulin preparations for insulin therapy. First, regular insulin is absorbed slowly by the subcutaneous tissue. After 30 to 60 minutes, the metabolic effect begins, and the highest concentration is reached after two to three hours of injection. As a result, people who take insulin regularly are more likely to experience postmeal hyperglycemia and late-postprandial hypoglycemia. Second, peak glucose is markedly reduced by the conventional basal insulin isophane (NPH). NPH is absorbed from subcutaneous tissue at varying rates [22]. Due to these pharmacodynamic limitations, users are more likely to experience hypoglycemia at night and elevated glucose levels before pre-breakfast. Insulin analogs based on a modified amino acid sequence from the human insulin molecule have been developed to address these issues.

Due to reduced self-association, the three fast-acting analogs (Lispro, Glulisin, and Aspart) are absorbed faster than regular insulin. Within 15 minutes of subcutaneous injection, they begin to work, and their peak effect is faster and stronger. The ideal non-spike duration of action for insulin glargine, the first long-acting insulin analog, was initially claimed to be nearly 24 hours [23]. However, these initial pharmacodynamic studies have been criticized, and it is safe to conclude that peakless insulin preparations do not exist [24, 25]. However, compared to NPH insulin, where glargine has a slightly longer action than detemir, both long-acting insulin analogs (glargine and detemir) have a limited peak effect and a longer mean duration of action [25].

Generally, people with T1DM typically need 50% basal insulin and 50% prandial insulin daily. However, this depends on several variables, such as whether the person consumes higher- or lower-carb meals. Weight can be used to estimate total daily insulin requirements, with typical doses ranging from 0.4 to 1.0 units/kg/day. During puberty, pregnancy, and medical conditions, greater amounts are needed [26].

Pre-meal use of shorter-acting insulins combined with longer-acting insulins is a typical multiple-dose regimen for people with T1DM. Titration of the long-acting starting dose is used to regulate fasting and overnight glucose levels. The timely injection of prandial insulin best controls postmeal glucose spikes. The optimal timing for administering prandial insulin differs depending on the pharmacokinetics of the formulation (regular, RAA, inhaled), pre-meal blood glucose levels, and carbohydrate intake. As a result, individual recommendations for prandial insulin administration should be made. Because physiological insulin secretion varies with glycemia, meal size, meal composition, and tissue demand for glucose, techniques have been developed to adjust prandial dosages based on anticipated demands to address this variability in individuals on insulin. Therefore, most people should be trained to adjust prandial insulin to their carbohydrate intake, pre-meal glucose levels, and anticipated activity [27, 28].

PHARMACOLOGIC THERAPY FOR ADULTS WITH TYPE 2 DIABETES

Considering comorbidities and treatment objectives, patient-centered treatment factors should guide pharmacologic therapy [29, 30]. Individualized factors influencing treatment choice include individualized blood glucose and weight goals, effects on weight, hypoglycemia, and cardiovascular protection; underlying physiological factors; drug side effect profiles, cost, and availability of medicines (Fig. **1**) [7].

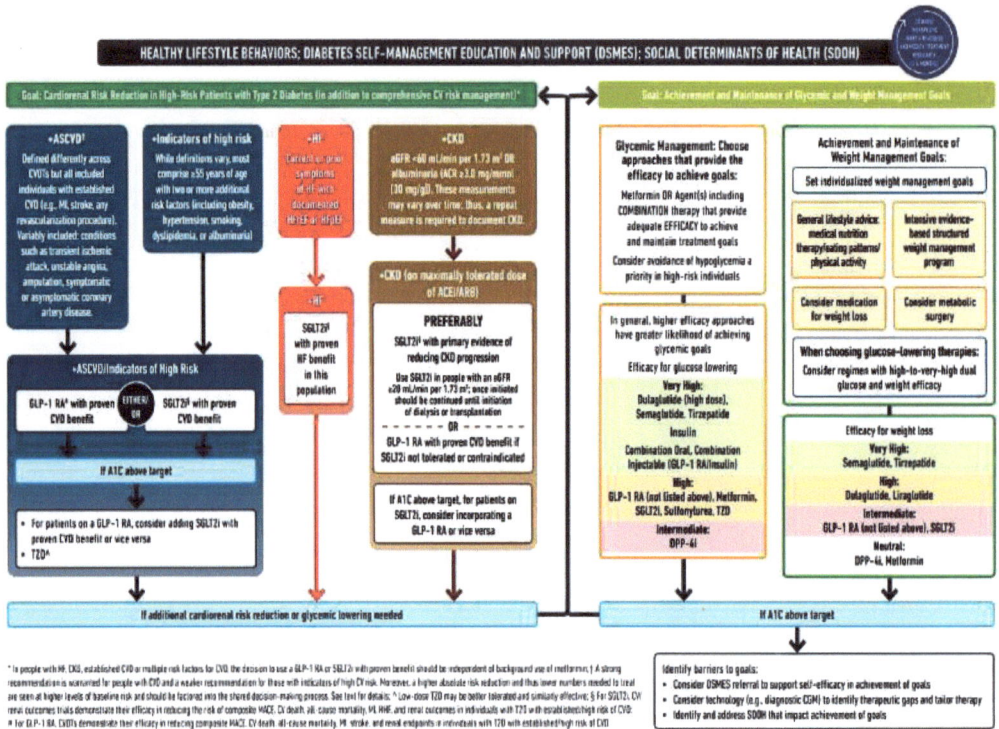

Fig. (1). 1—Use glucose-lowering medications to manage type 2 diabetes. ACEi, angiotensin-converting enzyme inhibitor; ACR, albumin-to-creatinine ratio; ARB, angiotensin receptor blocker; ASCVD, atherosclerotic cardiovascular disease; CGM, continuous glucose monitoring; CKD, chronic kidney disease; CV, cardiovascular; CVD, cardiovascular disease; CVOT, cardiovascular outcomes trial; DPP-4i, dipeptidyl peptidase 4 inhibitors; eGFR, estimated glomerular filtration rate; GLP-1 RA, glucagon-like peptide 1 receptor agonist; HF, heart failure; HFpEF, heart failure with preserved ejection fraction; HFrEF, heart failure with reduced ejection fraction; HHF, hospitalization for heart failure; MACE, major adverse cardiovascular events; MI, myocardial infarction; SDOH, social determinants of health; SGLT2i, sodium-glucose co-transporter 2 inhibitors; T2D, type 2 diabetes; TZD, thiazolidinedione [29].

NON-INSULIN TREATMENT FOR TYPE 2 DIABETES MELLITUS

Several non-insulin-based oral therapies have been suggested to treat T2DM (Fig. **2**).

Insulin Secretagogues

The sulfonylurea receptor (SUR) of the ATP-sensitive potassium channel on pancreatic β-cells is the target of this family of medications, which increases the pancreas' release of insulin [32].

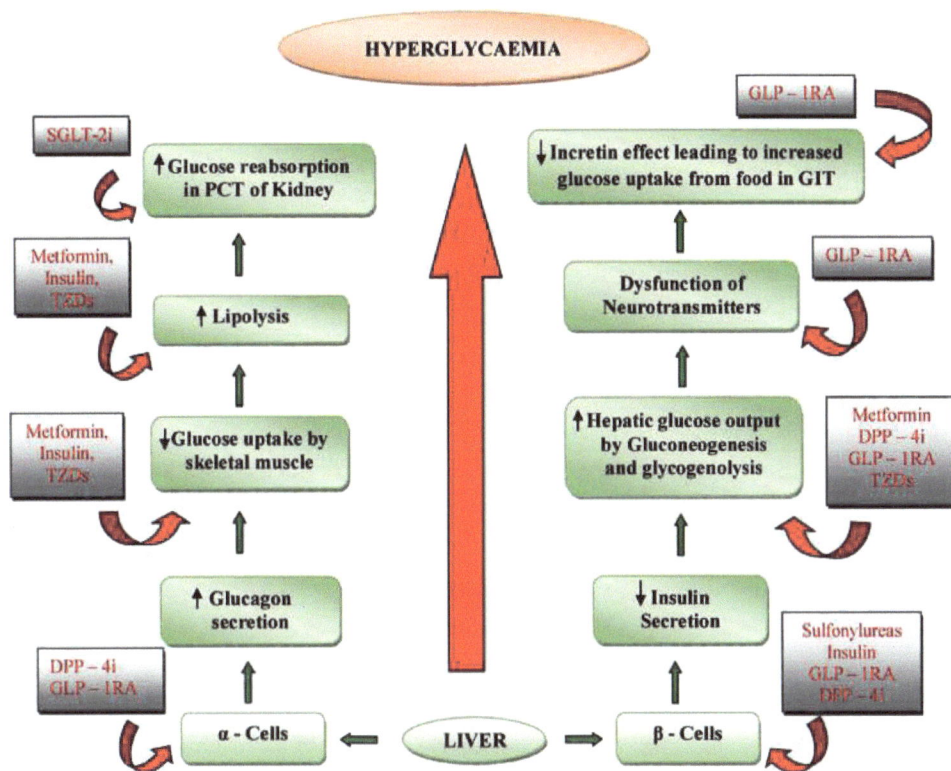

Fig. (2). Targets of treatment for T2DM [TZDs – Thiazolidinediones, DPP – 4i – Dipetidyl peptide – 4 inhibitor, GLP-1RA – Glucagon-like peptide – 1 receptor agonist, SGLT-2i - Sodium–Glucose co-transporter 2 inhibitor] [31].

Sulfonylureas

This class of medications works by attaching to the sulfonylurea receptor (SUR) of the ATP-sensitive potassium channel on pancreatic β-cells, boosting insulin production from the pancreas [32]. Sulfonylureas (SU) slow the liver's ability to clear insulin, prevent lipids from being broken down into fatty acids and restrict liver gluconeogenesis [33]. Tolbutamide, Chlorpropamide, Tolazamide, and Acetohexamide are sulfonylureas of the first generation. Sulfonylureas of the second generation, such as Glibenclamide, Glipizide, and Glimepiride, developed because of their increased potency, quicker onset of action, and shorter plasma half-lives [34]. Common side effects of sulfonylurea are hypoglycemia, weight gain, skin reactions, and upset stomach [35].

SUs have continued to be one of the backbones of pharmacotherapy for treating T2DM ever since they were first used in clinical practice in the 1950s. Although

SUs are safe, effective, and beneficial, newer therapies must still overshadow their clinical value and place in treatment. However, there is a wealth of information, experience, and, most importantly, outcome data supporting the use of modern SUs in the management of diabetic patients [36].

SUs are only effective when there is still activity in pancreatic β-cell because their primary function is enhancing insulin secretion. SU-caused hypoglycemia can last many hours, necessitating hospitalization every time [37]. In a randomized meta-analysis based on clinical trials, glimepiride was found to be as effective as metformin in controlling blood sugar level [38].

Hypoglycemia is the major concern of SUs. Long-acting tolbutamide and chlorpropamide are associated with greater hypoglycemia risks. Therefore, it does not apply to older people. Short-acting drugs such as gliclazide or glibenclamide are recommended as an alternative to long-acting drugs [20].

Meglitinides

Meglitinides are insulin secretagogues with a short half-life focusing on the early loss of prandial insulin secretion [39]. They depolarize the cell by closing potassium channels on the cell membrane dependent on adenosine triphosphate and acting glucose-dependently. This causes calcium channels to open, which increases calcium influx and insulin secretion [40, 41]. The recommended intake time is 15 minutes before each meal [41].

Although they bind to the SU receptor at a different location, the glinides stimulate insulin secretion in the same way SUs do [42]. They may have a decreased risk of hypoglycemia but have a comparable risk of weight gain as SUs. Repaglinide has the advantage of being able to be given in renal insufficiency because the active ingredient is mainly excreted by the liver. Repaglinide potency is comparable to the majority of other antihyperglycemics, whereas nateglinide appears to be less effective [43 - 45].

Amylin Analogs

Amylin is excreted from β-cells in the pancreas along with insulin [46]. It slows down the emptying of the stomach and reduces glucagon secretion, thereby maintaining both fasting and postprandial blood glucose levels. through regulating the brain's appetite center, it controls how much food one eats [47]. In both T1DM and T2DM, amylin deficiency exists. Because it is aggregated and insoluble in solution, amylin cannot be used as a drug, so its analogs capable of mimicking the action of amylin have been developed. Amylin analogs are available in a parenteral form [48]. The drug in this family that is now available is pramlintide acetate, which is given subcutaneously [48, 49]. Nausea, vomiting, headaches, and hypoglycemia after taking insulin are the most frequent adverse effects of amylin analogs. As the patient adapts to the medicine, these adverse effects disappear [48].

Biguanides

It decreases hepatic glucose production and intestinal glucose absorption and enhances insulin sensitivity [50]. Unlike insulin secretagogues, these molecules do not directly affect insulin secretion. Due to the high prevalence of associated lactic acidosis, phenformin and buformin have been removed from clinical use. Metformin, on the other hand, is widely used and has a much lower risk of lactic acidosis [51]. Biguanides have vasoprotective properties and anti-hypertriglyceridemia activity and do not result in hypoglycemia or weight gain. Biguanides activate AMP-dependent protein kinase, which prevents fatty acid breakdown [52]. Biguanides, on the other hand, frequently cause gastrointestinal discomfort, such as diarrhea, cramps, nausea, vomiting, and increased gas and bloating. Vitamin B12 absorption decreases over time with its use [53].

Metformin

Metformin is commonly used to treat T2DM. Phenformin was identified in extracts of the Galega officinalis (French lilac) plant in the 1920s, and isoamylene guanidine, also known as galegin, has been used to treat DM for centuries [54]. Metformin has been used in Europe since 1957 and in the United States since 1995 for its blood-sugar-lowering effects. Despite being the most used anti-diabetic medication administered globally, little is known about how it works [55]. Through its intricate interactions with mitochondrial enzymes, metformin induces the liver's adenosine monophosphate-activated protein kinase to become active, causing the liver to absorb glucose and inhibiting gluconeogenesis [56]. In addition, it increases tyrosine kinase activity and activates insulin receptor expression to enhance insulin sensitivity. Recent research reveals that metformin protects cardiovascular disease by lowering plasma lipid levels through a peroxisome proliferator-activated receptor (PPAR) pathway [56]. The reduction in food intake could be brought about by incretin-like actions mediated by glucagon-like peptide-1 (GLP-1). In people who are overweight or obese and at risk for diabetes, metformin may result in moderate weight reduction [57].

Metformin is effective and safe and comes in two dosage options: an immediate-release version for twice-day administration and an extended-release form for use once daily. First-line metformin has a better effect on HbA1C, weight, and cardiovascular mortality than sulfonylureas [58]. The main side effects of metformin are diarrhea, bloating, and gastrointestinal intolerance. Stepwise dose titration can help alleviate these symptoms. The drug is eliminated through renal filtration, and lactic acidosis has been linked to very high levels in the blood (for example, as a result of an overdose or acute renal failure). The FDA updated the metformin label to highlight its safety in individuals with an eGFR of less than 30 mL/min/1.73 m2. However, this issue is relatively uncommon, and metformin may still be taken safely in these patients [59]. As enough insulin is produced, metformin is quite effective, but when diabetes progresses to the point of β-cell failure, it loses its effectiveness [60].

Insulin Sensitizers

In the early 1990s, it was determined that the nuclear receptor superfamily 1 included the peroxisome proliferator-activated receptor (PPAR) family. It was discovered that thiazolidinediones (TZDs), known as "glitazones," have a strong inhibitory effect on PPAR-γ (60). These receptors belong to three subtypes, PPAR-α, δ, and γ. Specific to maintaining glucose homeostasis is PPAR. The effects of TZDs on PPAR-γ facilitated transcription and increased whole-body insulin sensitivity [61].

Two TZDs that can be used in a clinical setting at the moment are pioglitazone and rosiglitazone. One of the basic disease mechanisms of T2DM is insulin resistance in the target tissues; as insulin sensitizers, TZDs counteract this disease mechanism [62]. Systemic fatty acid production and uptake are also reduced by glitazones. Skeletal muscle glucose uptake is enhanced by PPAR-γ activation, which also delays gluconeogenesis to reduce glucose production [63].

These drugs are associated with common side effects such as edema, weight gain, macular edema, heart failure, lower hematocrit and hemoglobin levels, and increased risk of bone fractures [64]. Common side effects of these drugs include edema, weight gain, macular edema, heart failure, decreased hematocrit, hemoglobin levels, and an increased risk of bone fractures [64].

Dual PPAR-α/γ agonists have recently been found to have anti-diabetic properties. Maintaining lipid metabolism, insulin sensitivity, and inflammatory control are all maintained by activating PPAR-α and PPAR-γ receptors in tandem. PPAR-γ agonist side effects are lessened with dual therapy. Examples of dual PPAR agonists include muraglitazar, tesaglitazar, aleglitazar, ragaglitazar, naveglitazar, and saroglitazar [65]. Due to cardiotoxicity, muraglitazar has been drawn from the market [66]. It has been demonstrated that TZDs have beneficial effects on components of the "metabolic syndrome" and vascular reactivity indices [67, 68].

Rosiglitazone would be restricted by the FDA due to the possibility of an increased risk of cardiovascular complications. Moreover, a cohort study [69] found that exposure to pioglitazone may significantly contribute to the development of bladder cancer.

Alpha-glucosidase Inhibitors (AGIs)

In the 1990s, AGIs were approved for the treatment of T2DM [70]. Oral anti-diabetic medications called alpha-glucosidase inhibitors (AGIs) are particularly useful in T2DM with excessive postprandial glucose. By taking AGIs, the undigested carbohydrates moved to the distal section of the small intestine and large intestine, delaying the process of carbohydrate absorption in the gastrointestinal system [71]. AGIs are saccharides that slow the digestion of carbohydrates by acting as competitive inhibitors for enzymes in the small intestine (Fig. **3**). This slows the rate at which glucose from the diet enters the bloodstream and reduces postprandial hyperglycemia [72].

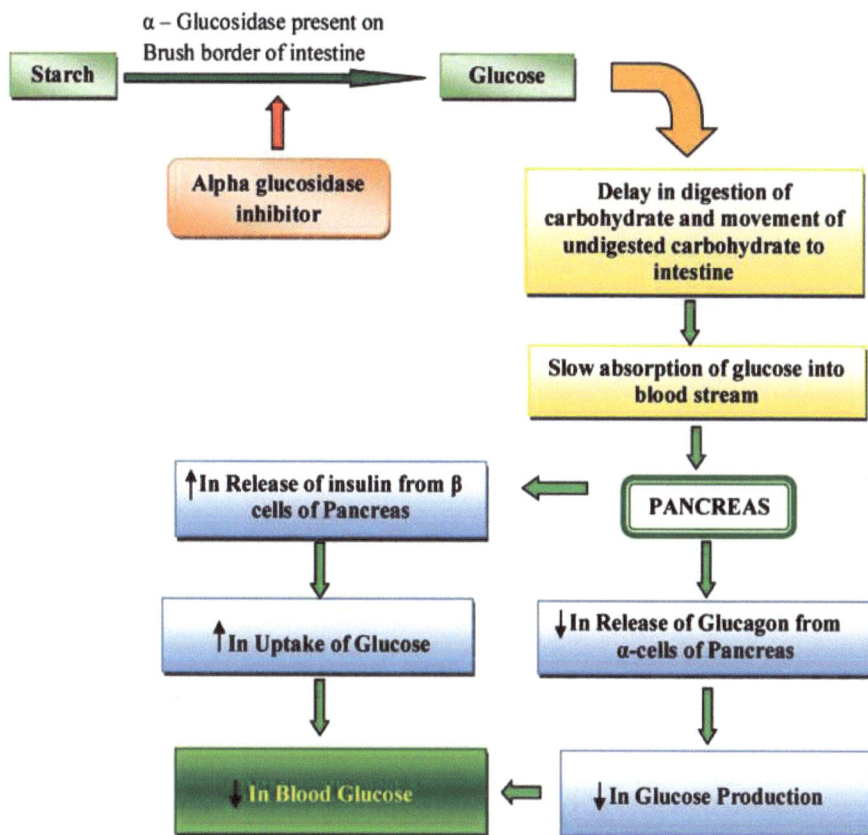

Fig. (3). The schematic mechanism for α-glucosidase inhibitors [31].

Currently approved as diabetes treatment are Acarbose, Miglitol, and Voglibose, and three members of this group [72, 73]. Gastrointestinal discomfort is a common AGI side effect, which can recover in a matter of weeks. Alpha-glucosidase inhibitors are not recommended for inflammatory bowel disease, intestinal obstruction, liver cirrhosis, and pregnant women [74, 75]. Acarbose has been linked to a few cases of hepatitis, which resolved when the drug was stopped; Consequently, liver enzymes ought to be evaluated prior to and during treatment with this medication [76].

Incretin Mimetics

The difference between the insulin secretory response to an oral glucose load and an intravenous glucose load is known as the incretin effect. Following oral glucose ingestion, the incretin effect accounts for 50–70% of all insulin production [77]. Glucagon-like peptide-1 (GLP-1) and glucose-dependent insulinotropic polypeptide (GIP), two naturally occurring incretin hormones crucial for maintaining glycemic control, have short half-lives due to DPP-4 inhibitors' fast hydrolysis of them within a minute. Incretin activity (especially GIP)is diminished or absent in T2DM patients. Incretins decrease gastric emptying and cause weight loss [57].

GLP-1 Agonists

A series of naturally occurring metabolic hormones (GLP-1 and GIP) known as incretins promote a drop in blood sugar levels [78]. After eating a meal, L-cells in the intestine secrete GLP-1, a peptide with 36 amino acids. The pancreatic beta cells produce and secrete insulin in response to GLP-1 [79]. The ATP-sensitive potassium channel closes, and the membrane depolarizes as a result of carbohydrate metabolism in the intestinal L-cells, leading to calcium ions (Ca^{2+}) entry. GLP-1 is secreted as a result of this [80]. Due to the rapid metabolism of GLP-1 by dipeptidyl peptide IV (DPP-IV) enzymes, its half-life is approximately one to two minutes [46]. Therefore, the demand for developing GLP-1 analogs with a longer half-life could be an avenue for treating diabetes mellitus. Once more, DPP-IV inhibitors function as incretin mimics [81]. The newest class of injectable medications used to treat T2DM are GLP-1 agonists or analogs [81 - 83]. GLP-1 analogs were stable and twice as potent as the original GLP-1 [84].

The first GLP-1 analog was Exenatide. It resembled human GLP-1 by 53% and was resistant to DPP-IV. It is taken twice daily and comes in a variety of brands. Lixisenatide is another GLP-1 analog that can be taken once per day. Liraglutide once daily and dulaglutide once weekly [85, 86]. Diarrhea, nausea, vomiting, headache, dizziness, increased sweating, indigestion, constipation, and loss of appetite are among the adverse effects of incretin mimetics [86]. Because they only stimulate insulin release and suppress glucagon secretion when blood glucose levels are elevated, GLP-1 receptor agonists are well-suited for early use in T2DM. This keeps the risk of hypoglycemia low [86] (Fig. **4**).

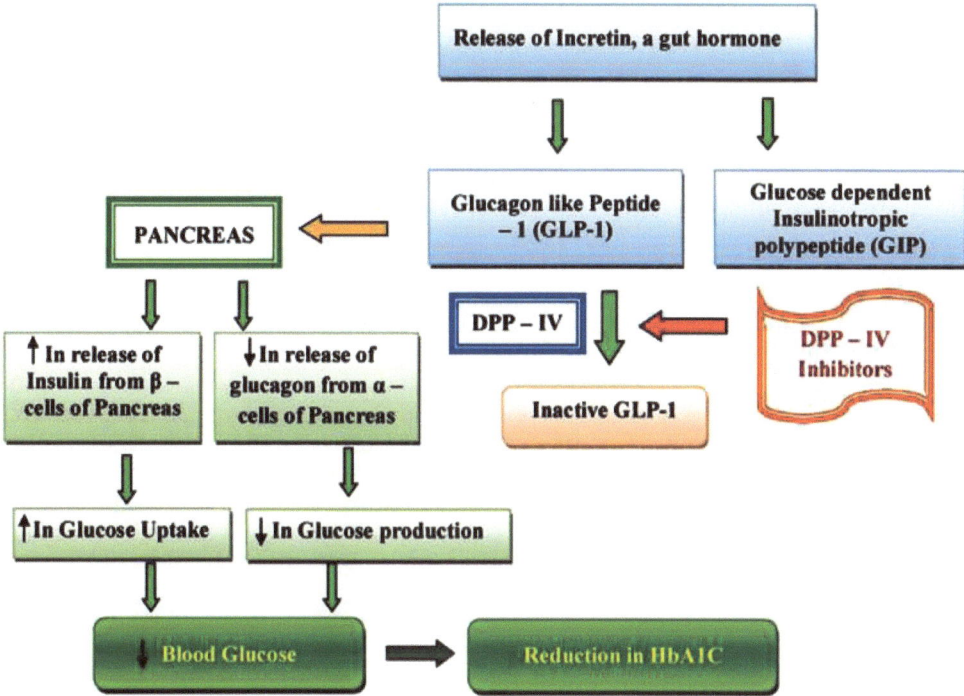

Fig. (4). The schematic mechanism of incretin mimetics [31].

Patients unable to achieve their HbA1C goals on metformin alone should take GLP-1 receptor agonists in combination with metformin as a dual therapy [87]. GLP-1 receptor agonists, metformin, and a sodium-glucose co-transporter 2 (SGLT-2) inhibitor can be combined in patients with persistent hyperglycemia who require triple therapy. Overweight patients benefit greatly from this triple combination [88]. In addition, using incretin with basal insulin may delay meal-related insulin by reducing hypoglycemia risk. The need to match mealtime insulin to specific carbohydrate ratios is reduced by this simplified regimen, which also helps prevent insulin use-induced weight gain [88]. Beyond glucose control, GLP-1 receptor agonists have emerged as powerful agents in cardiovascular prevention. They have demonstrated significant benefits in reducing major adverse cardiovascular events (MACE), including heart attack, stroke, and cardiovascular-related death. Furthermore, GLP-1 receptor agonists are associated with remarkable weight loss effects due to their ability to suppress appetite and slow down gastric emptying. In accordance with a study, certain GLP-1 receptor agonists have shown weight loss of approximately 4 to 6 kilograms over a treatment duration of around 6 to 12 months, underscoring their dual role in both cardiovascular risk reduction and weight management [89].

DPP-4 Inhibitors

Dipeptidyl peptidase-4 (DPP-4) is a serine protease in a membrane-bound and plasma-soluble form, which breaks down several biologically important peptides [90] (Fig. **5**). This class includes vildagliptin, sitagliptin, saxagliptin, linagliptin, dutogliptin, and alogliptin [91, 92].

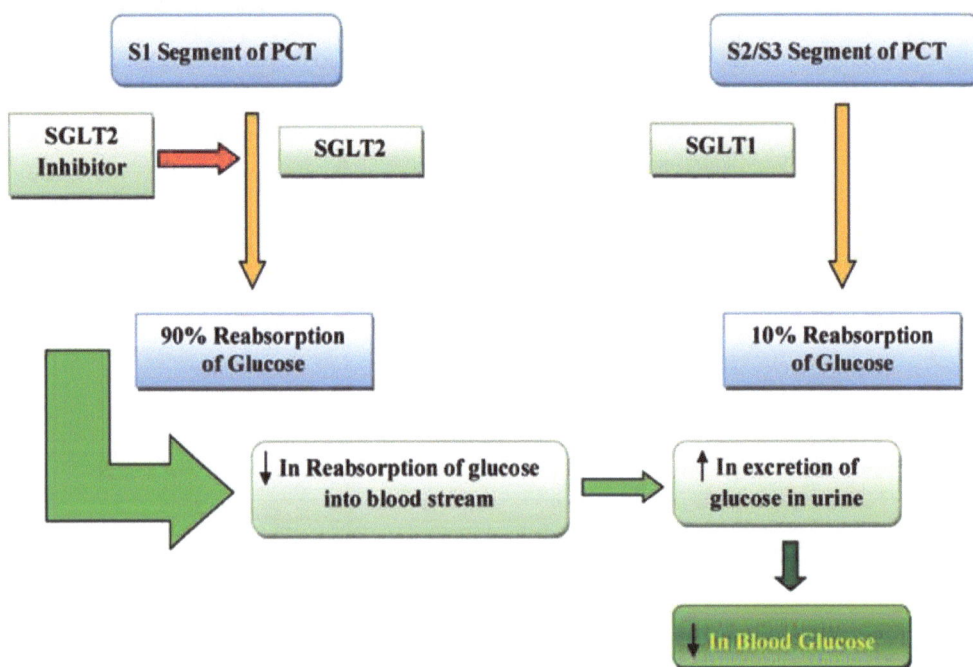

Fig. (5). The schematic mechanism of SGLT2 inhibitors action [31].

Through reversible inhibition of the DPP-4 enzyme, this class of drugs causes pancreatic cells to secrete less insulin, less glucagon, and less hepatic glucose [91].

Acceptable drug bioavailabilities were recorded in this group unaffected by food intake. In patients with T2DM who are not adequately controlled on monotherapy with the anti-diabetic medications that are currently available, these agents have also been effective as combination therapy [93]. The effectiveness of intensifying insulin therapy is outweighed by combining a DPP-4 inhibitor with basal insulin therapy [94]. When pioglitazone therapy demonstrated poor glycemic control in patients and sitagliptin was supplemented, it has also been reported that combination therapy with TZDs and DPP-4 inhibitors was more effective [95].

DPP-4 inhibitor therapy rarely results in hypoglycemia. DPP-4 inhibitors are safe and effective for the elderly, and the low risk of hypoglycemia makes them an excellent choice for this population of diabetics [96]. When a DPP-4 inhibitor is added to an SU or insulin, hypoglycemia is more likely to occur [97]. Compared to metformin and DPP-4 combination treatment, metformin alone has a much higher prevalence of gastrointestinal adverse effects [98]. Other reported adverse reactions were nasopharyngitis, upper respiratory tract infection, headache, and acute pancreatitis [57].

Body weight is not affected by DPP-4 inhibitor therapy [99]. It has been demonstrated that this class of drugs for renal impairment is safe and effective [100]. In patients with hepatic impairment, these drugs' pharmacokinetic profiles do not significantly alter; As a result, they are safe to use for liver impairment [101]. However, it has been reported that vildagliptin 100 mg taken once daily rarely causes hepatic dysfunction, and hepatic function should be monitored prior to and during treatment [102]. When taking a variety of DPP-4 inhibitors, there is no increased risk of cardiovascular or cerebrovascular side effects [103].

Sodium-glucose Co-transporter Two Inhibitors

The active co-transporter, the sodium-glucose co-transporter (SGLT), and the passive transporter, the facilitative glucose transporter (GLUT), are responsible for reabsorbing glucose in the proximal convoluted tubule (PCT) [104]. SGLT2 inhibitors prevent glucose reabsorption in PCT and boost glucose excretion in the urine (Fig. **5**), so blood glucose levels and other glycemic parameters are preserved [105]. Canagliflozin, dapagliflozin, empagliflozin, ipragliflozin, luseogliflozin, and tofogliflozin are the obtainable molecules that fall under this category [106]. SGLT2 inhibitors are prescribed as monotherapy or in combination with metformin, sulfonylureas, thiazolidinediones, or insulin [107, 108].

In a diabetic animal model, phlorizin, a natural compound, nonselectively inhibited SGLT1 and SGLT2, reversing insulin resistance and normalizing plasma glucose concentration. As a result, researchers became interested in SGLT2 inhibitors [109, 110]. SGLT2 inhibitors have demonstrated sustained clinical efficacy in post-marketing studies [111 - 113]. Without increasing the number of severe hypoglycemia episodes, SGLT2 inhibitors can be used alone or in conjunction with insulin to lower HbA1C levels, stabilize insulin doses, and reduce body weight [114]. Stable weight loss is increased when metformin and SGLT2 inhibitors are combined [115]. On the other hand, losing weight is suitable for controlling blood pressure and glucose levels [116]. SGLT2 inhibitors cause permanent induction of diastolic or systolic blood pressure suppression due to body weight reduction and sodium excretion enhancement [117, 118]. Lowering uric acid levels is another advantage of this therapy [118].

The risk of hypoglycemia is very low. However, using SGLT2 inhibitors in addition to SUs only reveals the increased frequency of hypoglycemic episodes [119]. Recent meta-analyses have indicated that SGLT2 inhibitors do not increase the risk of urinary tract infections, just confirming the prevalence of genital fungal infections, particularly in women [120]. Osmotic diuresis-associated side effects of this class of diabetes medications include an increase in hematocrit and a decrease in intravascular volume [121].

Orthostatic hypotension is an uncommon occurrence. However, it is advised that older persons who are more susceptible to dehydration use these medications with caution. In patients sensitive to volume changes, it is particularly wise to keep an eye out for the telltale indications of electrolyte imbalance and dehydration when SGLT2 inhibitors are given together with diuretics [122, 123]. SGLT2 inhibitor treatment has been linked to euglycemic ketoacidosis and latent autoimmune diabetes in some studies. Canagliflozin, an SGLT2 inhibitor, is also reported to lower bone mineral density and increase the risk of bone fractures and the rate of biochemical markers related to bone formation and resorption [122]. Numerous studies demonstrate that the use of SGLT2 inhibitors is not harmful to cardiovascular health. Recent research has illuminated the significant cardiovascular benefits of SGLT2 inhibitors. One study underscores SGLT2 inhibitors' role in improving cardiovascular outcomes, showcasing their potential to mitigate cardiac dysfunction and reduce the risk of adverse cardiovascular events. Additionally, another study contributes to this growing body of evidence, highlighting the positive impact of SGLT2 inhibitors on cardiac performance. A further layer of support is provided by another research that demonstrates the significant strides made in enhancing cardiovascular health through SGLT2 inhibitors administration. Gliflozins demonstrated a positive impact on cardiac remodeling, leading to significant enhancements in left ventricular systolic and diastolic function, as well as improvements in left atrial reservoir and total emptying function. Additionally, gliflozins were associated with a reduction in pulmonary artery pressure. Collectively, these studies underscore SGLT2 inhibitors' potential to not only optimize cardiovascular outcomes but also improve cardiac function, marking a promising avenue in the realm of cardiovascular therapy [123 - 125].

Insulin in Type 2 Diabetes

When catabolic features (weight loss, hypertriglyceridemia, ketosis) are present, insulin should be considered as part of any combination treatment for severe hyperglycemia because it can be effective where other agents are not. If a person has a blood glucose level of 300 mg/dL or an HbA1C of more than 10%, signs of catabolism (such as weight loss), or symptoms of hyperglycemia (such as polyuria or polydipsia), insulin therapy is typically started. When glucose toxicity subsides, it is often possible to simplify the regimen and switch to non-insulin agents [126].

Depending on the severity of the hyperglycemia, basal insulin (titrated by 2–3 U every 4–7 days until the glycemic target is reached) is the initial insulin treatment. If basal insulin is used at a concentration greater than 0.5 U/kg, an additional agent is required. Prandial insulin can be added to meals if basal insulin contributes to acceptable fasting blood glucose, but HbA1C remains consistently above target. The total insulin dose can be calculated, with half administered as a basal dose and the other half administered during meals, evenly distributed over three meals. Lispro, Aspart, or Glulisine are fast-acting insulin analogs that can be utilized and given just before meals. Before meals and after injections, glucose levels should be checked [57].

Another approach to controlling preprandial glucose spikes may be to add premixed (or biphasic) insulin analogs (70/30 aspart mix, 75/25, or 50/50 lispro mix) twice daily. Premixed human NPH/Regular (70/30) formulations and regular human insulin are more affordable substitutes for fast-acting and premixed insulin analogs, respectively. However, they are inadequate for addressing postmeal glucose fluctuations because of their unpredictably pharmacodynamic profiles [57].

A continuous subcutaneous insulin infusion pump, on the other hand, can be utilized to avoid requiring multiple injections. Currently, prandial usage of inhaled insulin is possible (Insulin delivery system Technosphere, Afrezza). The dosage range is constrained, nevertheless. A pulmonary function test is necessary both before and after the initiation of treatment using inhaled insulin. Patients with asthma or other lung diseases should not use it [57].

An obstacle to using insulin in T2DM may be weight gain. Patients taking sulfonylureas gained 1.7-2.6 kilograms, while those taking insulin gained 6 kilograms in the UKPDS [127]. In recent years, the use of insulin and GLP-1 receptor agonists in combination has effectively prevented insulin-induced weight gain and avoided the requirement for high doses in patients with significant insulin resistance.

CONCLUSION

In this chapter, different conventional medications for the treatment of diabetes are discussed and their mechanism of action, side effects, and safety profiles are presented. Diabetes drugs are given in monotherapy or combination and the significant challenges in effective diabetes management are optimizing current treatments to ensure optimal and stable glucose control with minimal side effects and reducing long-term complications of diabetes.

REFERENCES

[1]   Khursheed R, Singh SK, Wadhwa S, *et al.* Treatment strategies against diabetes: Success so far and challenges ahead. Eur J Pharmacol 2019; 862: 172625.
[http://dx.doi.org/10.1016/j.ejphar.2019.172625] [PMID: 31449807]

[2]   Wild S, Roglic G, Green A, Sicree R, King H. Global prevalence of diabetes: Estimates for the year 2000 and projections for 2030. Diabetes Care 2004; 27(5): 1047-53.
[http://dx.doi.org/10.2337/diacare.27.5.1047] [PMID: 15111519]

[3]   Wong CY, Al-Salami H, Dass CR. Potential of insulin nanoparticle formulations for oral delivery and diabetes treatment. J Control Release 2017; 264: 247-75.
[http://dx.doi.org/10.1016/j.jconrel.2017.09.003] [PMID: 28887133]

[4]   ElSayed NA, Aleppo G, Aroda VR, *et al.* 9. Pharmacologic approaches to glycemic treatment: Standards of Care in diabetes—2023. Diabetes Care 2023; 46 (Suppl. 1): S140-57.
[http://dx.doi.org/10.2337/dc23-S009] [PMID: 36507650]

[5]   Tan SY, Mei Wong JL, Sim YJ, *et al.* Type 1 and 2 diabetes mellitus: A review on current treatment approach and gene therapy as potential intervention. Diabetes Metab Syndr 2019; 13(1): 364-72.
[http://dx.doi.org/10.1016/j.dsx.2018.10.008] [PMID: 30641727]

[6]   Souto EB, Souto SB, Campos JR, *et al.* Nanoparticle delivery systems in the treatment of diabetes complications. Molecules 2019; 24(23): 4209.
[http://dx.doi.org/10.3390/molecules24234209] [PMID: 31756981]

[7]   ElSayed NA, Aleppo G, Aroda VR, *et al.* 9. Pharmacologic Approaches to Glycemic Treatment: *Standards of Care in Diabetes—2023*. Diabetes Care 2023; 46 (Suppl. 1): S140-57.
[http://dx.doi.org/10.2337/dc23-S009] [PMID: 36507650]

[8]   Cleary PA, Orchard TJ, Genuth S, *et al.* The effect of intensive glycemic treatment on coronary artery calcification in type 1 diabetic participants of the Diabetes Control and Complications Trial/Epidemiology of Diabetes Interventions and Complications (DCCT/EDIC) Study. Diabetes 2006; 55(12): 3556-65.
[http://dx.doi.org/10.2337/db06-0653] [PMID: 17130504]

[9]   Nathan DM, Cleary PA, Backlund JY, *et al.* Diabetes control and complications trial/epidemiology of diabetes interventions and complications (dcct/edic) study research group intensive diabetes treatment and cardiovascular disease in patients with type 1 diabetes. N Engl J Med 2005; 353(25): 2643-53.
[http://dx.doi.org/10.1056/NEJMoa052187] [PMID: 16371630]

[10]   Diabetes Control and Complications Trial/Epidemiology of Diabetes Interventions and Complications (DCCT/EDIC) Study Research Group Diabetes Care 2016; 13(19): 1378-83.

[11]   Holt RIG, DeVries JH, Hess-Fischl A, *et al.* The Management of Type 1 Diabetes in Adults. A Consensus Report by the American Diabetes Association (ADA) and the European Association for the Study of Diabetes (EASD). Diabetes Care 2021; 44(11): 2589-625.
[http://dx.doi.org/10.2337/dci21-0043] [PMID: 34593612]

[12]   Tricco AC, Ashoor HM, Antony J, *et al.* Safety, effectiveness, and cost effectiveness of long acting versus intermediate acting insulin for patients with type 1 diabetes: systematic review and network meta-analysis. BMJ 2014; 349(oct01 6): g5459.
[http://dx.doi.org/10.1136/bmj.g5459] [PMID: 25274009]

[13]   Bartley PC, Bogoev M, Larsen J, Philotheou A. Long-term efficacy and safety of insulin detemir compared to Neutral Protamine Hagedorn insulin in patients with Type 1 diabetes using a treat-t--target basal–bolus regimen with insulin aspart at meals: A 2-year, randomized, controlled trial. Diabet Med 2008; 25(4): 442-9.
[http://dx.doi.org/10.1111/j.1464-5491.2007.02407.x] [PMID: 18387078]

[14]   DeWitt DE, Hirsch IB. Outpatient insulin therapy in type 1 and type 2 diabetes mellitus: scientific review. JAMA 2003; 289(17): 2254-64.
[http://dx.doi.org/10.1001/jama.289.17.2254] [PMID: 12734137]

[15]   Bode BW, McGill JB, Lorber DL, Gross JL, Chang PC, Bregman DB. Inhaled technosphere insulin compared with injected prandial insulin in type 1 diabetes: A randomized 24-week trial. Diabetes Care 2015; 38(12): 2266-73.
[http://dx.doi.org/10.2337/dc15-0075] [PMID: 26180109]

[16]   Russell-Jones D, Bode BW, De Block C, *et al.* Fast-Acting Insulin Aspart Improves Glycemic Control in Basal-Bolus Treatment for Type 1 Diabetes: Results of a 26-Week Multicenter, Active-Controlled, Treat-to-Target, Randomized, Parallel-Group Trial (onset 1). Diabetes Care 2017; 40(7): 943-50.https://care.diabetesjournals.org/content/40/7/943
[http://dx.doi.org/10.2337/dc16-1771] [PMID: 28356319]

[17]   Blevins T, Zhang Q, Frias JP, Jinnouchi H, Chang AM. Randomized double-blind clinical trial comparing ultra rapid lispro with lispro in a basal-bolus regimen in patients with type 2 diabetes: PRONTO-T2D. Diabetes Care 2020; 43(12): 2991-8.
[http://dx.doi.org/10.2337/dc19-2550] [PMID: 32616612]

[18]   Lane W, Bailey TS, Gerety G, *et al.* Effect of Insulin Degludec vs Insulin Glargine U100 on Hypoglycemia in Patients With Type 1 Diabetes. JAMA 2017; 318(1): 33-44.
[http://dx.doi.org/10.1001/jama.2017.7115] [PMID: 28672316]

[19]   Home PD, Bergenstal RM, Bolli GB, *et al.* New Insulin Glargine 300 Units/mL Versus Glargine 100 Units/mL in People With Type 1 Diabetes: A Randomized, Phase 3a, Open-Label Clinical Trial (EDITION 4). Diabetes Care 2015; 38(12): 2217-25.
[http://dx.doi.org/10.2337/dc15-0249] [PMID: 26084341]

[20]   Pervin R. Current Anti-diabetic Drugs. Nutritional and Therapeutic Interventions for Diabetes and Metabolic Syndrome 2018; pp. 455-73.
[http://dx.doi.org/10.1016/B978-0-12-812019-4.00034-9]

[21]   Melander A. Oral antidiabetic drugs: An overview. Diabet Med 1996; 13(9) (Suppl. 6): 143-7.
[http://dx.doi.org/10.1002/dme.1996.13.s6.143] [PMID: 8894498]

[22]   Holleman F, Gale EAM. Nice insulins, pity about the evidence. Diabetologia 2007; 50(9): 1783-90.
[http://dx.doi.org/10.1007/s00125-007-0763-4] [PMID: 17634918]

[23]   Lepore M, Pampanelli S, Fanelli C, *et al.* Pharmacokinetics and pharmacodynamics of subcutaneous injection of long-acting human insulin analog glargine, NPH insulin, and ultralente human insulin and continuous subcutaneous infusion of insulin lispro. Diabetes 2000; 49(12): 2142-8.
[http://dx.doi.org/10.2337/diabetes.49.12.2142] [PMID: 11118018]

[24]   Heise T, Pieber TR. Towards peakless, reproducible and long-acting insulins. An assessment of the basal analogues based on isoglycaemic clamp studies. Diabetes Obes Metab 2007; 9(5): 648-59.
[http://dx.doi.org/10.1111/j.1463-1326.2007.00756.x] [PMID: 17645556]

[25]   Porcellati F, Rossetti P, Busciantella NR, *et al.* Comparison of pharmacokinetics and dynamics of the long-acting insulin analogs glargine and detemir at steady state in type 1 diabetes: A double-blind,

randomized, crossover study. Diabetes Care 2007; 30(10): 2447-52.
[http://dx.doi.org/10.2337/dc07-0002] [PMID: 17623819]

[26]   Peters AL, Laffel L, Chiang JL. American Diabetes Association/JDRF Type 1 diabetes sourcebook. Alexandria, Virginia: American Diabetes Association 2013.

[27]   Bell KJ, Barclay AW, Petocz P, Colagiuri S, Brand-Miller JC. Efficacy of carbohydrate counting in type 1 diabetes: A systematic review and meta-analysis. Lancet Diabetes Endocrinol 2014; 2(2): 133-40.
[PMID: 24622717]

[28]   Vaz EC, Porfírio GJM, Nunes HRC, Nunes-Nogueira VS. Effectiveness and safety of carbohydrate counting in the management of adult patients with type 1 diabetes mellitus: A systematic review and meta-analysis. Arch Endocrinol Metab 2018; 62(3): 337-45.
[http://dx.doi.org/10.20945/2359-3997000000045] [PMID: 29791661]

[29]   Davies MJ, Aroda VR, Collins BS, *et al.* Management of hyperglycemia in Type 2 diabetes, 2022. A consensus Report by the American Diabetes Association (ADA) and the european association for the study of diabetes (EASD). Diabetes Care 2022; 45(11): 2753-86.
[http://dx.doi.org/10.2337/dci22-0034] [PMID: 36148880]

[30]   ElSayed NA, Aleppo G, Aroda VR, *et al.* 2. Classification and Diagnosis of Diabetes: *Standards of Care in Diabetes—2023.* Diabetes Care 2023; 46 (Suppl. 1): S19-40.
[http://dx.doi.org/10.2337/dc23-S002] [PMID: 36507649]

[31]   Padhi S, Nayak AK, Behera A. Type II diabetes mellitus: A review on recent drug based therapeutics. Biomed Pharmacother 2020; 131: 110708.
[http://dx.doi.org/10.1016/j.biopha.2020.110708] [PMID: 32927252]

[32]   Rastegari A, Rabbani M, Sadeghi HM, Imani EF, Hasanzadeh A, Moazen F. Pharmacogenetic association of kcnj11 (e23k) variant with therapeutic response to sulphonylurea (glibenclamide) in iranian patients. Int J Diabetes Dev Ctries 2015; 35(4): 630-1.
[http://dx.doi.org/10.1007/s13410-015-0316-1]

[33]   Proks P, Reimann F, Green N, Gribble F, Ashcroft F. Sulfonylurea stimulation of insulin secretion. Diabetes 2002; 51 (Suppl. 3): S368-76.
[http://dx.doi.org/10.2337/diabetes.51.2007.S368] [PMID: 12475777]

[34]   Kalra S, Bahendeka S, Sahay R, *et al.* Consensus recommendations on sulfonylurea and sulfonylurea combinations in the management of Type 2 diabetes mellitus – International Task Force. Indian J Endocrinol Metab 2018; 22(1): 132-57.
[http://dx.doi.org/10.4103/ijem.IJEM_556_17] [PMID: 29535952]

[35]   Sola D, Rossi L, Schianca GPC, *et al.* State of the art paper Sulfonylureas and their use in clinical practice. Arch Med Sci 2015; 4(4): 840-8.
[http://dx.doi.org/10.5114/aoms.2015.53304] [PMID: 26322096]

[36]   Inzucchi SE, Bergenstal RM, Buse JB, *et al.* Management of hyperglycemia in type 2 diabetes: A patient-centered approach: position statement of the American Diabetes Association (ADA) and the European Association for the Study of Diabetes (EASD). Diabetes Care 2012; 35(6): 1364-79.
[http://dx.doi.org/10.2337/dc12-0413] [PMID: 22517736]

[37]   Sarkar A, Tiwari A, Bashin P, Mitra M. Pharmacological and pharmaceutical profile of gliclazide: A review. J Appl Pharm Sci 2011; 01(09): 11-9.

[38]   Zhu H, Zhu S, Zhang X, *et al.* Comparative efficacy of glimepiride and metformin in monotherapy of type 2 diabetes mellitus: meta-analysis of randomized controlled trials. Diabetol Metab Syndr 2013; 5(1): 70.
[http://dx.doi.org/10.1186/1758-5996-5-70] [PMID: 24228743]

[39]   Polonsky KS, Given BD, Hirsch LJ, *et al.* Abnormal patterns of insulin secretion in non-insuli-
-dependent diabetes mellitus. N Engl J Med 1988; 318(19): 1231-9.

[http://dx.doi.org/10.1056/NEJM198805123181903] [PMID: 3283554]

[40]   Gromada J, Dissing S, Kofod H, Frøkjaer-Jensen J. Effects of the hypoglycaemic drugs repaglinide and glibenclamide on ATP-sensitive potassium-channels and cytosolic calcium levels in beta TC3 cells and rat pancreatic beta cells. Diabetologia 1995; 38(9): 1025-32.
[http://dx.doi.org/10.1007/BF00402171] [PMID: 8591815]

[41]   Owens DR, Luzio SD, Ismail I, Bayer T. Increased prandial insulin secretion after administration of a single preprandial oral dose of repaglinide in patients with type 2 diabetes. Diabetes Care 2000; 23(4): 518-23.
[http://dx.doi.org/10.2337/diacare.23.4.518] [PMID: 10857945]

[42]   Malaisse WJ. Pharmacology of the meglitinide analogs: New treatment options for type 2 diabetes mellitus. Treat Endocrinol 2003; 2(6): 401-14.
[http://dx.doi.org/10.2165/00024677-200302060-00004] [PMID: 15981944]

[43]   Derosa G, Mugellini A, Ciccarelli L, Crescenzi G, Fogari R. Comparison between repaglinide and glimepiride in patients with type 2 diabetes mellitus: A one-year, randomized, double-blind assessment of metabolic parameters and cardiovascular risk factors. Clin Ther 2003; 25(2): 472-84.
[http://dx.doi.org/10.1016/S0149-2918(03)80090-5] [PMID: 12749508]

[44]   Madsbad S, Kilhovd B, Lager I, Mustajoki P, Dejgaard A. Comparison between repaglinide and glipizide in Type 2 diabetes mellitus: A 1-year multicentre study. Diabet Med 2001; 18(5): 395-401.
[http://dx.doi.org/10.1046/j.1464-5491.2001.00490.x] [PMID: 11472451]

[45]   Saloranta C, Hershon K, Ball M, Dickinson S, Holmes D. Efficacy and safety of nateglinide in type 2 diabetic patients with modest fasting hyperglycemia. J Clin Endocrinol Metab 2002; 87(9): 4171-6.
[http://dx.doi.org/10.1210/jc.2002-020068] [PMID: 12213867]

[46]   Rang HP, Dale MM, Ritter JM, Moore P. Pharmacology. Churchill Livingstone 2003.

[47]   Adeghate E, Kalász H. Amylin analogues in the treatment of diabetes mellitus: medicinal chemistry and structural basis of its function. Open Med Chem J 2011; 5 (Suppl. 2): 78-81.
[http://dx.doi.org/10.2174/1874104501105010078] [PMID: 21966328]

[48]   Schmitz O, Brock B, Rungby J. Amylin agonists: A novel approach in the treatment of diabetes. Diabetes 2004; 53 (Suppl. 3): S233-8.
[http://dx.doi.org/10.2337/diabetes.53.suppl_3.S233] [PMID: 15561917]

[49]   Hoogwerf B, Doshi KB, Diab D. Pramlintide, the synthetic analogue of amylin: Physiology, pathophysiology, and effects on glycemic control, body weight, and selected biomarkers of vascular risk. Vasc Health Risk Manag 2008; 4(2): 355-62.
[http://dx.doi.org/10.2147/VHRM.S1978] [PMID: 18561511]

[50]   Quillen DM, Samraj G, Kuritzky L. Improving management of type 2 diabetes mellitus: 2. Biguanides. Hosp Pract 1999; 34(11): 41-4.
[http://dx.doi.org/10.1080/21548331.1999.11443925] [PMID: 10887428]

[51]   Rubiño , Carrillo E, Alcalá , Domínguez-Martín A, Marchal , Boulaiz H. Phenformin as an Anticancer Agent: Challenges and Prospects. Int J Mol Sci 2019; 20(13): 3316.
[http://dx.doi.org/10.3390/ijms20133316] [PMID: 31284513]

[52]   Bourron O, Daval M, Hainault I, *et al.* Biguanides and thiazolidinediones inhibit stimulated lipolysis in human adipocytes through activation of AMP-activated protein kinase. Diabetologia 2010; 53(4): 768-78.
[http://dx.doi.org/10.1007/s00125-009-1639-6] [PMID: 20043143]

[53]   Sanchez-Rangel E, Inzucchi SE. Metformin: Clinical use in type 2 diabetes. Diabetologia 2017; 60(9): 1586-93.
[http://dx.doi.org/10.1007/s00125-017-4336-x] [PMID: 28770321]

[54]   Bailey CJ, Day C. Traditional plant medicines as treatments for diabetes. Diabetes Care 1989; 12(8): 553-64.

[http://dx.doi.org/10.2337/diacare.12.8.553] [PMID: 2673695]

[55]   Liu C, Wu D, Zheng X, Li P, Li L. Efficacy and safety of metformin for patients with type 1 diabetes mellitus: A meta-analysis. Diabetes Technol Ther 2015; 17(2): 142-8.
[http://dx.doi.org/10.1089/dia.2014.0190] [PMID: 25369141]

[56]   Viollet B, Guigas B, Garcia NS, Leclerc J, Foretz M, Andreelli F. Cellular and molecular mechanisms of metformin: An overview. Clin Sci (Lond) 2012; 122(6): 253-70.
[http://dx.doi.org/10.1042/CS20110386] [PMID: 22117616]

[57]   Chaudhury A, Duvoor C, Reddy Dendi VS, Kraleti S, Chada A, Ravilla R, *et al.* Clinical review of anti-diabetic drugs: Implications for type 2 diabetes mellitus management. Front Endocrinol 2017; 8(6).

[58]   Maruthur NM, Tseng E, Hutfless S, *et al.* Diabetes Medications as Monotherapy or Metformin-Based Combination Therapy for Type 2 Diabetes. Ann Intern Med 2016; 164(11): 740-51.
[http://dx.doi.org/10.7326/M15-2650] [PMID: 27088241]

[59]   Food US. FDA Drug Safety Communication: FDA revises warnings regarding use of the diabetes medicine metformin in certain patients with reduced kidney function . Drug Administration 2016.

[60]   Bailey CJ, Krentz AJ. Oral antidiabetic agents. InTextof diabe 2010; 452-77.
[http://dx.doi.org/10.1002/9781444324808.ch29]

[61]   Thangavel N, Al Bratty M, Akhtar Javed S, Ahsan W, Alhazmi HA. Targeting peroxisome proliferator-activated receptors using thiazolidinediones: strategy for design of novel antidiabetic drugs. Int J Med Chem 2017; 2017: 1-20.
[http://dx.doi.org/10.1155/2017/1069718] [PMID: 28656106]

[62]   Goldstein BJ. Differentiating members of the thiazolidinedione class: A focus on efficacy. Diabetes Metab Res Rev 2002; 18(S2) (Suppl. 2): S16-22.
[http://dx.doi.org/10.1002/dmrr.251] [PMID: 11921434]

[63]   Greenfield JR, Chisholm DJ. Experimental and clinical pharmacology: Thiazolidinediones - mechanisms of action. Aust Prescr 2004; 27(3): 67-70.
[http://dx.doi.org/10.18773/austprescr.2004.059]

[64]   Lebovitz HE. Thiazolidinediones: the Forgotten Diabetes Medications. Curr Diab Rep 2019; 19(12): 151.
[http://dx.doi.org/10.1007/s11892-019-1270-y] [PMID: 31776781]

[65]   Balakumar P, Mahadevan N, Sambathkumar R. A Contemporary Overview of PPARα/γ Dual Agonists for the Management of Diabetic Dyslipidemia. Curr Mol Pharmacol 2019; 12(3): 195-201.
[http://dx.doi.org/10.2174/1874467212666190111165015] [PMID: 30636619]

[66]   Amato AA, de Assis Rocha Neves F. Idealized PPAR-Based Therapies: Lessons from Bench and Bedside. PPAR Res 2012; 2012: 1-9.
[http://dx.doi.org/10.1155/2012/978687] [PMID: 22745632]

[67]   McGuire DK, Inzucchi SE. New drugs for the treatment of diabetes mellitus: part I: Thiazolidinediones and their evolving cardiovascular implications. Circulation 2008; 117(3): 440-9.
[http://dx.doi.org/10.1161/CIRCULATIONAHA.107.704080] [PMID: 18212301]

[68]   Dormandy JA, Charbonnel B, Eckland DJA, *et al.* Secondary prevention of macrovascular events in patients with type 2 diabetes in the PROactive Study (PROspective pioglitAzone Clinical Trial In macroVascular Events): A randomised controlled trial. Lancet 2005; 366(9493): 1279-89.
[http://dx.doi.org/10.1016/S0140-6736(05)67528-9] [PMID: 16214598]

[69]   Park H, Park C, Kim Y, Rascati KL. Efficacy and safety of dipeptidyl peptidase-4 inhibitors in type 2 diabetes: meta-analysis. Ann Pharmacother 2012; 46(11): 1453-69.
[http://dx.doi.org/10.1345/aph.1R041] [PMID: 23136353]

[70]   Bischoff H. Pharmacology of alpha-glucosidase inhibition. Eur J Clin Invest 1994; 24 (Suppl. 3): 3-10.

[PMID: 8001624]

[71] Lee A, Patrick P, Wishart J, Horowitz M, Morley JE. The effects of miglitol on glucagon-like peptide-1 secretion and appetite sensations in obese type 2 diabetics. Diabetes Obes Metab 2002; 4(5): 329-35.
[http://dx.doi.org/10.1046/j.1463-1326.2002.00219.x] [PMID: 12190996]

[72] Wehmeier UF, Piepersberg W. Biotechnology and molecular biology of the? -glucosidase inhibitor acarbose. Appl Microbiol Biotechnol 2004; 63(6): 613-25.
[http://dx.doi.org/10.1007/s00253-003-1477-2] [PMID: 14669056]

[73] Narita T, Yokoyama H, Yamashita R, *et al.* Comparisons of the effects of 12-week administration of miglitol and voglibose on the responses of plasma incretins after a mixed meal in Japanese type 2 diabetic patients. Diabetes Obes Metab 2012; 14(3): 283-7.
[http://dx.doi.org/10.1111/j.1463-1326.2011.01526.x] [PMID: 22051162]

[74] Derosa G, Maffioli P. Mini-Special Issue paper Management of diabetic patients with hypoglycemic agents α-Glucosidase inhibitors and their use in clinical practice. Arch Med Sci 2012; 5(5): 899-906.
[http://dx.doi.org/10.5114/aoms.2012.31621] [PMID: 23185202]

[75] Holt RIG, Lambert KD. The use of oral hypoglycaemic agents in pregnancy. Diabet Med 2014; 31(3): 282-91.
[http://dx.doi.org/10.1111/dme.12376] [PMID: 24528229]

[76] Derosa G, Maffioli P. Mini-Special Issue paper Management of diabetic patients with hypoglycemic agents α-Glucosidase inhibitors and their use in clinical practice. Arch Med Sci 2012; 5(5): 899-906.https://www.ncbi.nlm.nih.gov/pmc/articles/PMC3506243/
[http://dx.doi.org/10.5114/aoms.2012.31621] [PMID: 23185202]

[77] Nauck MA, Meier JJ. The incretin effect in healthy individuals and those with type 2 diabetes: Physiology, pathophysiology, and response to therapeutic interventions. Lancet Diabetes Endocrinol 2016; 4(6): 525-36.
[http://dx.doi.org/10.1016/S2213-8587(15)00482-9] [PMID: 26876794]

[78] Knop FK, Hansen KB, Knop FK. Incretin mimetics: A novel therapeutic option for patients with type 2 diabetes – a review. Diabetes Metab Syndr Obes 2010; 3: 155-63.
[http://dx.doi.org/10.2147/DMSO.S7004] [PMID: 21437085]

[79] Sun EW, de Fontgalland D, Rabbitt P, *et al.* Mechanisms Controlling Glucose-Induced GLP-1 Secretion in Human Small Intestine. Diabetes 2017; 66(8): 2144-9.
[http://dx.doi.org/10.2337/db17-0058] [PMID: 28385801]

[80] MacDonald PE, El-kholy W, Riedel MJ, Salapatek AMF, Light PE, Wheeler MB. The multiple actions of GLP-1 on the process of glucose-stimulated insulin secretion. Diabetes 2002; 51 (Suppl. 3): S434-42.
[http://dx.doi.org/10.2337/diabetes.51.2007.S434] [PMID: 12475787]

[81] Holst JJ. Which to choose, an oral or an injectable glucagon-like peptide-1 receptor agonist? Lancet 2019; 394(10192): 4-6.
[http://dx.doi.org/10.1016/S0140-6736(19)31350-9] [PMID: 31186119]

[82] Lebovitz HE, Banerji MA. Non-insulin injectable treatments (glucagon-like peptide-1 and its analogs) and cardiovascular disease. Diabetes Technol Ther 2012; 14(S1) (Suppl. 1): S-43-50.
[http://dx.doi.org/10.1089/dia.2012.0022] [PMID: 22650224]

[83] Alexopoulos AS, Buse JB. Initial injectable therapy in type 2 diabetes: Key considerations when choosing between glucagon-like peptide 1 receptor agonists and insulin. Metabolism 2019; 98: 104-11.
[http://dx.doi.org/10.1016/j.metabol.2019.06.012] [PMID: 31255662]

[84] Gupta V. Glucagon-like peptide-1 analogues: An overview. Indian J Endocrinol Metab 2013; 17(3): 413-21.
[http://dx.doi.org/10.4103/2230-8210.111625] [PMID: 23869296]

[85] Manandhar B, Ahn JM. Glucagon-like peptide-1 (GLP-1) analogs: Recent advances, new possibilities,

and therapeutic implications. J Med Chem 2015; 58(3): 1020-37.
[http://dx.doi.org/10.1021/jm500810s] [PMID: 25349901]

[86] Yu M, Benjamin MM, Srinivasan S, *et al.* Battle of GLP-1 delivery technologies. Adv Drug Deliv Rev 2018; 130: 113-30.
[http://dx.doi.org/10.1016/j.addr.2018.07.009] [PMID: 30009885]

[87] St Onge E, Miller S, Clements E, Celauro L, Barnes K. The role of glucagon-like peptide-1 receptor agonists in the treatment of type 2 diabetes. J Transl Int Med 2017; 5(2): 79-89.
[http://dx.doi.org/10.1515/jtim-2017-0015] [PMID: 28721339]

[88] Vosoughi K, Atieh J, Khanna L, *et al.* Association of glucagon-like peptide 1 analogs and agonists administered for obesity with weight loss and adverse events: A systematic review and network meta-analysis. EClinicalMedicine 2021; 42: 101213.
[http://dx.doi.org/10.1016/j.eclinm.2021.101213] [PMID: 34877513]

[89] DeFronzo RA. Combination therapy with GLP-1 receptor agonist and SGLT2 inhibitor. Diabetes Obes Metab 2017; 19(10): 1353-62.
[http://dx.doi.org/10.1111/dom.12982] [PMID: 28432726]

[90] Lambeir AM, Durinx C, Scharpé S, De Meester I. Dipeptidyl-peptidase IV from bench to bedside: An update on structural properties, functions, and clinical aspects of the enzyme DPP IV. Crit Rev Clin Lab Sci 2003; 40(3): 209-94.
[http://dx.doi.org/10.1080/713609354] [PMID: 12892317]

[91] Deacon CF, Holst JJ. Dipeptidyl peptidase-4 inhibitors for the treatment of type 2 diabetes: comparison, efficacy and safety. Expert Opin Pharmacother 2013; 14(15): 2047-58.
[http://dx.doi.org/10.1517/14656566.2013.824966] [PMID: 23919507]

[92] Monami M, Dicembrini I, Mannucci E. Dipeptidyl peptidase-4 inhibitors and heart failure: A meta-analysis of randomized clinical trials. Nutr Metab Cardiovasc Dis 2014; 24(7): 689-97.
[http://dx.doi.org/10.1016/j.numecd.2014.01.017] [PMID: 24793580]

[93] Hollander P. A Review of Type 2 Diabetes Drug Classes. US Endocrinol 2008; 4(1): 58.
[http://dx.doi.org/10.17925/USE.2008.04.01.58]

[94] Frandsen CSS, Madsbad S. Efficacy and safety of dipeptidyl peptidase-4 inhibitors as an add-on to insulin treatment in patients with Type 2 diabetes: A review. Diabet Med 2014; 31(11): 1293-300.
[http://dx.doi.org/10.1111/dme.12561] [PMID: 25112609]

[95] Rosenstock J, Brazg R, Andryuk PJ, Lu K, Stein P. Efficacy and safety of the dipeptidyl peptidase-4 inhibitor sitagliptin added to ongoing pioglitazone therapy in patients with type 2 diabetes: A 24-week, multicenter, randomized, double-blind, placebo-controlled, parallel-group study. Clin Ther 2006; 28(10): 1556-68.
[http://dx.doi.org/10.1016/j.clinthera.2006.10.007] [PMID: 17157112]

[96] Barzilai N, Guo H, Mahoney EM, *et al.* Efficacy and tolerability of sitagliptin monotherapy in elderly patients with type 2 diabetes: A randomized, double-blind, placebo-controlled trial. Curr Med Res Opin 2011; 27(5): 1049-58.
[http://dx.doi.org/10.1185/03007995.2011.568059] [PMID: 21428727]

[97] Vilsbøll T, Rosenstock J, Yki-Järvinen H, *et al.* Efficacy and safety of sitagliptin when added to insulin therapy in patients with type 2 diabetes. Diabetes Obes Metab 2010; 12(2): 167-77.
[http://dx.doi.org/10.1111/j.1463-1326.2009.01173.x] [PMID: 20092585]

[98] Goldstein BJ, Feinglos MN, Lunceford JK, Johnson J, Williams-Herman DE. Effect of initial combination therapy with sitagliptin, a dipeptidyl peptidase-4 inhibitor, and metformin on glycemic control in patients with type 2 diabetes. Diabetes Care 2007; 30(8): 1979-87.
[http://dx.doi.org/10.2337/dc07-0627] [PMID: 17485570]

[99] Aschner P, Kipnes MS, Lunceford JK, Sanchez M, Mickel C, Williams-Herman DE. Effect of the dipeptidyl peptidase-4 inhibitor sitagliptin as monotherapy on glycemic control in patients with type 2

diabetes. Diabetes Care 2006; 29(12): 2632-7.
[http://dx.doi.org/10.2337/dc06-0703] [PMID: 17130196]

[100] Nowicki M, Rychlik I, Haller H, *et al.* Long-term treatment with the dipeptidyl peptidase-4 inhibitor saxagliptin in patients with type 2 diabetes mellitus and renal impairment: A randomised controlled 52-week efficacy and safety study. Int J Clin Pract 2011; 65(12): 1230-9.
[http://dx.doi.org/10.1111/j.1742-1241.2011.02812.x] [PMID: 21977965]

[101] Giorda CB, Nada E, Tartaglino B. Pharmacokinetics, safety, and efficacy of DPP-4 inhibitors and GLP-1 receptor agonists in patients with type 2 diabetes mellitus and renal or hepatic impairment. A systematic review of the literature. Endocrine 2014; 46(3): 406-19.
[http://dx.doi.org/10.1007/s12020-014-0179-0] [PMID: 24510630]

[102] Profit L, Chrisp P, Nadin C. Vildagliptin: The evidence for its place in the treatment of type 2 diabetes mellitus. Core Evid 2008; 3(1): 13-30.
[http://dx.doi.org/10.3355/ce.2008.009] [PMID: 20694081]

[103] Johansen O, Neubacher D, von Eynatten M, Patel S, Woerle HJ. Cardiovascular safety with linagliptin in patients with type 2 diabetes mellitus: A pre-specified, prospective, and adjudicated meta-analysis of a phase 3 programme. Cardiovasc Diabetol 2012; 11(1): 3.
[http://dx.doi.org/10.1186/1475-2840-11-3] [PMID: 22234149]

[104] Hsia DS, Grove O, Cefalu WT. An update on sodium-glucose co-transporter-2 inhibitors for the treatment of diabetes mellitus. Curr Opin Endocrinol Diabetes Obes 2016; 24(1): 1.
[http://dx.doi.org/10.1097/MED.0000000000000311] [PMID: 27898586]

[105] Scheen AJ. Pharmacodynamics, efficacy and safety of sodium-glucose co-transporter type 2 (SGLT2) inhibitors for the treatment of type 2 diabetes mellitus. Drugs 2015; 75(1): 33-59.
[http://dx.doi.org/10.1007/s40265-014-0337-y] [PMID: 25488697]

[106] Tentolouris A, Vlachakis P, Tzeravini E, Eleftheriadou I, Tentolouris N. SGLT2 inhibitors: A review of their antidiabetic and cardioprotective effects. Int J Environ Res Public Health 2019; 16(16): 2965.
[http://dx.doi.org/10.3390/ijerph16162965] [PMID: 31426529]

[107] Kalra S, Kesavadev J, Chadha M, Kumar GV. Sodium-glucose cotransporter-2 inhibitors in combination with other glucose-lowering agents for the treatment of type 2 diabetes mellitus. Indian J Endocrinol Metab 2018; 22(6): 827-36.
[http://dx.doi.org/10.4103/ijem.IJEM_162_17] [PMID: 30766826]

[108] Bakris GL, Fonseca VA, Sharma K, Wright EM. Renal sodium–glucose transport: Role in diabetes mellitus and potential clinical implications. Kidney Int 2009; 75(12): 1272-7.
[http://dx.doi.org/10.1038/ki.2009.87] [PMID: 19357717]

[109] Ehrenkranz JRL, Lewis NG, Ronald Kahn C, Roth J. Phlorizin: A review. Diabetes Metab Res Rev 2005; 21(1): 31-8.
[http://dx.doi.org/10.1002/dmrr.532] [PMID: 15624123]

[110] Kahn BB, Shulman GI, DeFronzo RA, Cushman SW, Rossetti L. Normalization of blood glucose in diabetic rats with phlorizin treatment reverses insulin-resistant glucose transport in adipose cells without restoring glucose transporter gene expression. J Clin Invest 1991; 87(2): 561-70.
[http://dx.doi.org/10.1172/JCI115031] [PMID: 1991839]

[111] Del Prato S, Nauck M, Durán-Garcia S, *et al.* Long-term glycaemic response and tolerability of dapagliflozin *versus* a sulphonylurea as add-on therapy to metformin in patients with type 2 diabetes: 4-year data. Diabetes Obes Metab 2015; 17(6): 581-90.
[http://dx.doi.org/10.1111/dom.12459] [PMID: 25735400]

[112] Wilding JPH, Woo V, Rohwedder K, Sugg J, Parikh S, Hoppichler F, *et al.* Dapagliflozin in patients with type 2 diabetes receiving high doses of insulin: efficacy and safety over 2 years. Diabetes Obes Metab 2014; 16(2): 124-36.
[http://dx.doi.org/10.1111/dom.12187] [PMID: 23911013]

[113] Ridderstråle M, Andersen KR, Zeller C, Kim G, Woerle HJ, Broedl UC. Comparison of empagliflozin and glimepiride as add-on to metformin in patients with type 2 diabetes: A 104-week randomised, active-controlled, double-blind, phase 3 trial. Lancet Diabetes Endocrinol 2014; 2(9): 691-700.
[http://dx.doi.org/10.1016/S2213-8587(14)70120-2] [PMID: 24948511]

[114] Ahmann A. Combination therapy in type 2 diabetes mellitus: Adding empagliflozin to basal insulin. Drugs Context 2015; 4(4): 1-7.
[http://dx.doi.org/10.7573/dic.212288] [PMID: 26633984]

[115] Bailey CJ, Morales Villegas EC, Woo V, Tang W, Ptaszynska A, List JF. Efficacy and safety of dapagliflozin monotherapy in people with Type 2 diabetes: A randomized double-blind placebo-controlled 102-week trial. Diabet Med 2015; 32(4): 531-41.
[http://dx.doi.org/10.1111/dme.12624] [PMID: 25381876]

[116] Sjöström CD, Hashemi M, Sugg J, Ptaszynska A, Johnsson E. Dapagliflozin-induced weight loss affects 24-week glycated haemoglobin and blood pressure levels. Diabetes Obes Metab 2015; 17(8): 809-12.
[http://dx.doi.org/10.1111/dom.12500] [PMID: 25997813]

[117] Cefalu WT, Stenlöf K, Leiter LA, *et al.* Effects of canagliflozin on body weight and relationship to HbA1c and blood pressure changes in patients with type 2 diabetes. Diabetologia 2015; 58(6): 1183-7.
[http://dx.doi.org/10.1007/s00125-015-3547-2] [PMID: 25813214]

[118] Cherney DZI, Perkins BA, Soleymanlou N, *et al.* Renal hemodynamic effect of sodium-glucose cotransporter 2 inhibition in patients with type 1 diabetes mellitus. Circulation 2014; 129(5): 587-97.
[http://dx.doi.org/10.1161/CIRCULATIONAHA.113.005081] [PMID: 24334175]

[119] Goring S, Hawkins N, Wygant G, *et al.* Dapagliflozin compared with other oral anti-diabetes treatments when added to metformin monotherapy: A systematic review and network meta-analysis. Diabetes Obes Metab 2014; 16(5): 433-42.
[http://dx.doi.org/10.1111/dom.12239] [PMID: 24237939]

[120] Liakos A, Karagiannis T, Athanasiadou E, *et al.* Efficacy and safety of empagliflozin for type 2 diabetes: A systematic review and meta-analysis. Diabetes Obes Metab 2014; 16(10): 984-93.
[http://dx.doi.org/10.1111/dom.12307] [PMID: 24766495]

[121] Lambers Heerspink HJ, de Zeeuw D, Wie L, Leslie B, List J. Dapagliflozin a glucose-regulating drug with diuretic properties in subjects with type 2 diabetes. Diabetes Obes Metab 2013; 15(9): 853-62.
[http://dx.doi.org/10.1111/dom.12127] [PMID: 23668478]

[122] Taylor SI, Blau JE, Rother KI. Possible adverse effects of SGLT2 inhibitors on bone. Lancet Diabetes Endocrinol 2015; 3(1): 8-10.
[http://dx.doi.org/10.1016/S2213-8587(14)70227-X] [PMID: 25523498]

[123] Cesaro A, Acerbo V, Vetrano E, *et al.* Sodium–glucose cotransporter 2 inhibitors in patients with diabetes and coronary artery disease: Translating the benefits of the molecular mechanisms of gliflozins into clinical practice. Int J Mol Sci 2023; 24(9): 8099.
[http://dx.doi.org/10.3390/ijms24098099] [PMID: 37175805]

[124] Palmiero G, Cesaro A, Galiero R, *et al.* Impact of gliflozins on cardiac remodeling in patients with type 2 diabetes mellitus & reduced ejection fraction heart failure: A pilot prospective study. GLISCAR study. Diabetes Res Clin Pract 2023; 200: 110686.
[http://dx.doi.org/10.1016/j.diabres.2023.110686] [PMID: 37100231]

[125] Russo V, D'Aquino MM, Caturano A, *et al.* Improvement of global longitudinal strain and myocardial work in type 2 diabetes patients on sodium-glucose cotransporter 2 inhibitors therapy. J Cardiovasc Pharmacol 2022; 10-97.
[http://dx.doi.org/10.1097/FJC.0000000000001450] [PMID: 37405837]

[126] Babu A, Mehta A, Guerrero P, *et al.* Safe and simple emergency department discharge therapy for patients with type 2 diabetes mellitus and severe hyperglycemia. Endocr Pract 2009; 15(7): 696-704.

[http://dx.doi.org/10.4158/EP09117.ORR] [PMID: 19625243]

[127] Uk prospective diabetes study (ukpds) group effect of intensive blood-glucose control with metformin on complications in overweight patients with type 2 diabetes (ukpds 34). Lancet 1998; 352(9131): 854-65.
[http://dx.doi.org/10.1016/S0140-6736(98)07037-8] [PMID: 9742977]

CHAPTER 3

Nanomedicine for Insulin Delivery in Diabetes

Morteza Rafiee-Tehrani[1], Somayeh Handali[1], Mohammad Vaziri[1], Sepideh Nezhadi[1] and Farid Abedin Dorkoosh[1,*]

[1] *Faculty of Pharmacy, Tehran University of Medical Sciences, Tehran, Iran*

Abstract: Diabetes is one of the common diseases in the world and its treatment faces challenges. Insulin is the main therapeutic agent used in the treatment of diabetic patients. However, it has several side effects and during the day, patients may need several insulin injections, which is not pleasant for them. Therefore, a controlled and prolonged release system is required to decrease the injection frequency, improve the bioavailability of insulin, and enhance the compliance of patients. Nanoparticles (NPs) based drug delivery systems (DDSs) have been considered for insulin delivery. NPs can improve the permeability of insulin by opening the tight junctions between intestinal epithelial cells and can protect insulin from the action of enzymes existing in the gastrointestinal (GI) tract.

Keywords: Delivery, Diabetes, Insulin, Nanoparticle.

INTRODUCTION

Diabetes is of two types in which the body cannot produce insulin (type 1) or is not sensitive to insulin (type 2) and; consequently, the blood glucose level is not well controlled [1]. Diabetes is one of the biggest global health challenges of the 21st century, and there are many people living with this disease [2, 3]. Diabetes is mainly caused by environmental influences, immune system dysfunction, mental factors, and genetics, which result in either insulin resistance or insufficient insulin secretion [4].

As mentioned, diabetes is a result of the inefficiency of insulin to convert glucose into energy. When this process is disrupted, blood glucose can rise to a level that has health consequences. Insulin is the main therapeutic agent used in the treatment of patients with diabetes [5]. Insulin is a globular protein with a molecular weight of 5808 Daltons, containing two chains, A (21 residues) and B

* **Corresponding author Farid Abedin Dorkoosh:** Faculty of Pharmacy, Tehran University of Medical Sciences, Tehran, Iran; E-mail: dorkoosh@tums.ac.ir

Ali Rastegari (Ed.)

(30 residues) that are linked together by disulfide bonds (Fig. **1**) [6]. Insulin is produced by β cells of the pancreas. When there is too much glucose in the blood, insulin converts extra glucose into glycogen and stores it in the liver [7].

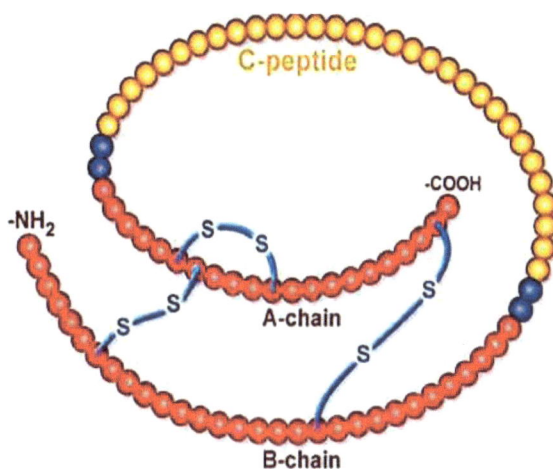

Fig. (1). Structure of insulin [8].

At present, the route of insulin administration is subcutaneous injection [9] which has several disadvantages including; lipohyperatrophy, obesity, retinopathy, hypoglycemia, neuropathy, lipoatrophy, allergic reactions and peripheral hyperinsulinemia [4, 10]. During the day, patients may need several insulin injections, which is not pleasant for most patients. Therefore, a controlled and prolonged release system is required to decrease the injection frequency and enhance the compliance of patients. According to the reports of Health Care Costs Institute, the cost of insulin for patients has doubled. Generally, oral delivery of peptide drugs is attractive due to patient compliance, patient adherence, and a cost-effective manufacturing process than injections. However, the gastrointestinal (GI) tract is a hostile milieu for the oral absorption of drugs due to low pH, existence of peptidases and proteases, and poor absorption through the intestinal epithelial layer. Furthermore, the hydrophilic nature of peptide drugs and the large molecular size further restrict their oral absorption [5, 11]. The bioavailability of insulin following oral delivery is usually lower than 1% owing to the enzymes existing in the GI tract and poor absorption *via* the intestinal epithelial cells [5].

Nanotechnology is a novel technology that will promote the next industrial revolution. Nanoparticles (NPs) based drug delivery systems (DDSs) have been considered to effectively transport various therapeutic agents to target cells. Nanocarriers are used for entrapment of drugs to limit their side effects and improve their bioavailability. Organic and inorganic NPs have been employed for drug delivery. Various nanocarriers such as drug delivery system such as liposomes, polymeric NPs, solid lipid NPs (SLNs), chitosan, exosomes, micelles, nanogels and dendrimers have been widely investigated for encapsulation of insulin and increasing its bioavailability (Fig. **2**). The structure of some nanocarriers for insulin delivery is shown in Fig. (**3**). These nanocarriers are biodegradable, biocompatible, non-toxic and can escape from the reticuloendothelial system. Insulin encapsulated in NPs can be protected from the action of the enzymes existing in the GI tract. Moreover, NPs can improve the permeability of insulin to the intestinal mucosa through opening the tight junctions between intestinal epithelial cells [12]. In recent years, long-acting NPs formulations containing insulin have been developed to diminish the frequency of injections. Additionally, nanocarriers are widely considered for oral delivery of insulin [13]. Smart nanocarrier-based drug delivery systems were also developed for insulin delivery. For example, glucose-responsive NPs (synthesized from dextran) were prepared for rapid and extended self-regulated insulin delivery. Results showed that these formulations could reduce the elevated blood glucose levels in mice and decrease the risk of hypoglycemia [1]. Glucose-responsive self-assembled polyamines as smart NPs were also used for insulin delivery. These smart NPs could appropriately regulate blood glucose concentration [14].

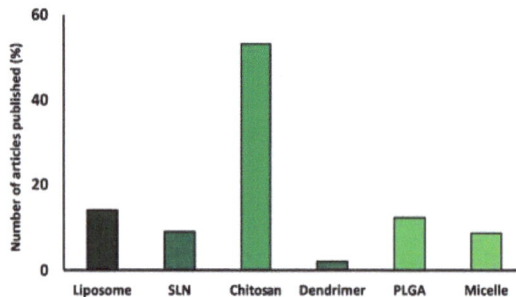

Fig. (2). Number of articles published in Elsevier, Springer, ACS, and Taylor & Francis journals from 2015 to 2022.

In oral insulin delivery, nanocarriers can improve the transport of insulin through the paracellular pathways. Chitosan can increase the paracellular transport of insulin through the interaction of positively charged polymers with the negatively charged cell membrane. Transcellular pathway is another mechanism for transporting insulin-loaded NPs. The transcellular pathway includes fusion, endocytosis, and adsorption. Moreover, receptor-mediated endocytosis is the major route for insulin-loaded nanocarriers to enter into cells [6, 15]. Fig. (**4**) illustrates the different delivery pathways of the insulin-loaded NPs.

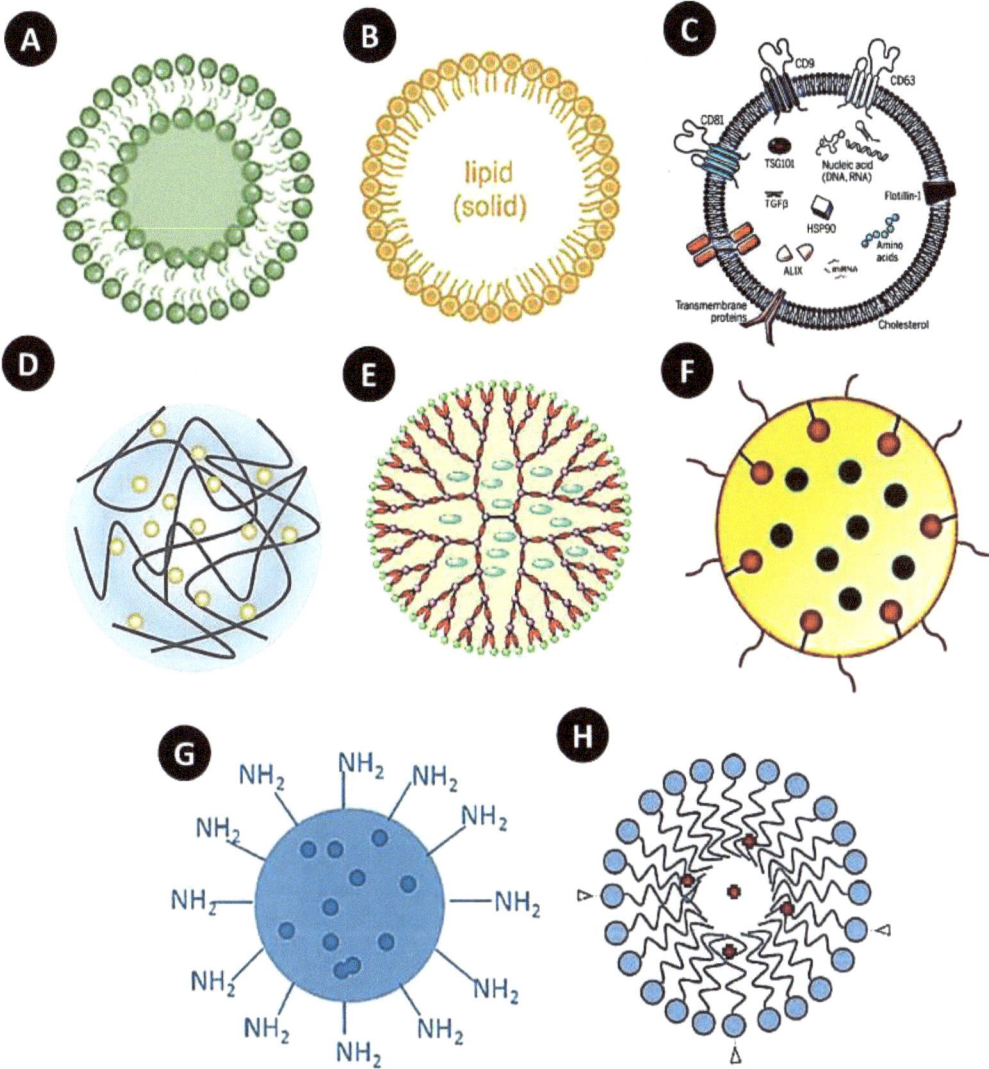

Fig. (3). Different types of nanocarriers for insulin delivery: **A**) liposome, **B**) SLN, **C**) exosome, **D**) nanogel, **E**) dendrimer, **F**) PLGA NPs **G**) chitosan NPs and **H**) polymeric micelle.

NANOCARRIERS BASED INSULIN DELIVERY

Due to the drawbacks of conventional injectable insulin, different nanocarriers have been developed for efficient delivery of insulin.

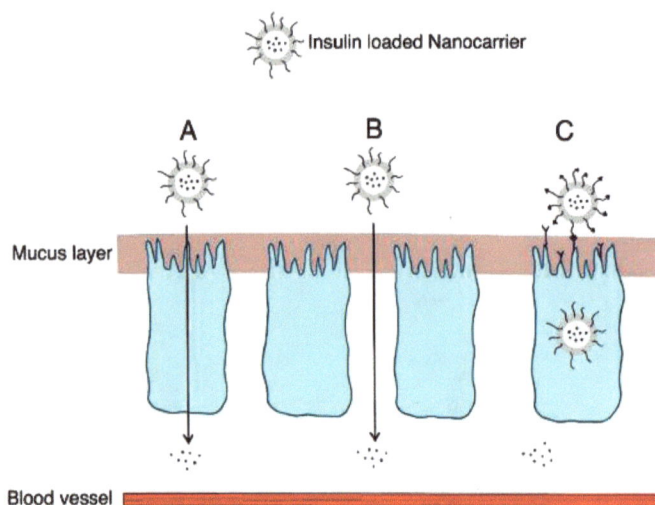

Fig. (4). Schematic illustration of the different delivery pathways for insulin-loaded NPs: **A**) transcellular transport, **B**) paracellular transport, and **C**) receptor-mediated endocytosis.

Liposome

Liposomes, as spherical vesicles are composed of lipid bilayers that protect drugs from rapid elimination and degradation. Liposomes are biocompatible, biodegradable, non-toxic and have the ability to entrap hydrophilic and hydrophobic drugs [4]. Liposomes can deliver drugs through various ways, including buccal, oral, nasal and transdermal routes [16]. Encapsulation of insulin into liposomes leads to increasing its stability owing to resistance against enzymatic activity [13]. However, conventional liposomes are instable in the GI tract and have poor permeability across the intestinal epithelial cells. Modification of liposomes leads to improving their stability *via* the GI tract and increasing their uptake through endocytosis pathway [5]. Liposomes functionalized with cell-penetrating peptides (CPPs) were designed for insulin delivery through nasal route. According to the results, insulin-loaded liposomes functionalized with CPPs promoted the release, permeation, and permeability coefficient of insulin through the nasal mucosa more than liposomal formulations without functionalization [17]. Moreover, encapsulation of insulin into liposomes led to increasing its stability owing to resistance against enzymatic activity [13]. Thiamine and niacin decorated liposomes were developed for enhancing oral delivery of insulin. Results of *in vivo* showed that liposomal formulation improved the interaction of insulin with gastrointestinal vitamin receptors and increased its oral absorption [13]. Targeting the folate receptors in the intestine can improve the absorption of nanocarriers in the GI tract [18]. Folate targeted PEGylated liposomes reduced blood glucose and increased insulin level in mice.

It was concluded that folate receptors mediated endocytosis which enhanced the bioavailability and cellular uptake of insulin [5]. Folic acid is a necessary nutrient for its receptors that are present on the GI tract cells; therefore, conjugation of folic acid with nanocarriers can effectively improve their uptake through folate receptors mediated endocytosis [12]. Insulin presents good wound-healing properties to restore the integrity of the broken skin. A liposomal chitosan gel was developed to deliver insulin. The stability study showed liposomal formulation was stable over 6 months in an aqueous dispersion and displayed a sustained release. Daily application of these liposomes in clinical studies presented promising benefits which over control. These findings indicated the effectiveness of the formulation to improve wound healing [19]. Sodium-glycodeoxycholate (SGDC)-incorporated elastic liposomes were characterized for insulin permeation across porcine buccal tissues. According to the findings, this nanoformulation remarkably increased the permeation of insulin across buccal tissues [16].

Chitosan

Chitosan, as a natural polymeric polysaccharide is a linear polysaccharide that composed of β-(1-4)- linked d-glucosamine and N-acetyl-d-glucosamine [10]. Chitosan shows favorable properties such as biocompatibility, biodegradability, non-immunogenicity, ability to bind with organic compounds, and non-toxicity and it also enhances permeation of drugs [20, 21]. Due to the presence of -NH2 and -OH groups, chitosan has a unique mucoadhesive property [22]. It has been reported that chitosan NPs can promote the permeation of loaded insulin *via* paracellular absorption and endocytic pathway, open up the tight junction between epithelial cells, and enhance the glucose uptake in cells [10, 20]. Folate conjugated chitosan NPs increased the stability of insulin to the harsh GI tract and enhanced its cellular uptake. Moreover, insulin-loaded folate chitosan NPs improved hypoglycemic activity in *in vivo* [18]. However, chitosan shows low stability in the acidic milieu owing to its extreme cationic charge [23]. In a study, poly (sodium 4-styrenesulfonate, PSS) was used as a cross-linking for improving the stability of chitosan NPs in acidic media and folic acid was also employed for increasing cellular uptake. Results showed that insulin-loaded folic acid-chitosan NPs could induce hypoglycemia and increase the bioavailability and stability of insulin more than insulin-loaded chitosan NPs without modification [23]. A carboxymethyl-b-cyclodextrin-grafted chitosan NPs (insulin/CMCD-g-CS NPs) was designed for oral insulin delivery. These nanocarriers exhibited excellent encapsulation efficiency, sustained drug release profile, and enhanced drug penetration into cells through opening the tight junction. Moreover, these synthesized NPs significantly reduced blood glucose level than the mice and did not display any obvious toxic effects on experimental animals [7]. A thermosensitive copolymer incorporating chitosan-zinc-insulin complex was designed for insulin delivery. Thermosensitive triblock copolymer was poly (D, L-lactide)-poly (ethylene glycol)-poly (D, L-lactide) (PLA-PEG-PLA) which is biodegradable, biocompatible and dissolves easily in water; consequently, the use of toxic organic solvents in the formulation is avoided. Zinc has a vital role in the synthesis and storage of insulin in β cells of the pancreatic. Modification of insulin with zinc and chitosan incorporated into a thermosensitive copolymeric system reduced the burst release and improved its stability during storage from denaturation [24].

PLGA

Polymeric NPs are extensively considered as drug delivery systems for drug delivery. Poly (lactic-co-glycolic) acid (PLGA) is mostly used as a drug delivery system due to its biodegradability and biocompatibility features and is also approved by FDA for numerous medical applications [25, 26]. PLGA NPs can enhance the absorption of insulin by promoting the transcellular pathway of insulin. Furthermore, it has been shown that insulin encapsulated PLGA NPs reduced serum glucose levels *in vivo*, which confirms the improved oral absorption of insulin using a polymeric nanocarriers. In the study, PLGA/chitosan NPs were developed for insulin delivery. These nanocarriers showed significantly hypoglycemic effect [25]. A montmorillonite PLGA nanocomposite was evaluated for oral delivery of insulin. These carriers increased the stability of insulin in the environment with low pH and showed a sustained release profile for insulin in the simulated gastrointestinal fluid [27].

Solid Lipid Nanoparticle (SLN)

SLNs are composed of a solid lipid core and a layer of surfactants [28]. SLNs are non-toxic, biodegradable, have high tolerability and extended blood residence time, provide controlled and sustained release, lead to increased bioavailability, and can be easily produced at the industrial scale [10, 29, 30]. VB12-coated gel-core-SLN was designed for oral delivery of insulin. This formulation prohibited the burst release of insulin from the nanocarrier in the low pH environment; consequently, avoided the rapid enzymatic degradation. The insulin encapsulated in VB12-coated gel-core-SLN also exhibited a prolonged hypoglycemic effect in diabetic rats [31]. A cationic SLN (cSLN) was developed for oral delivery of insulin. Cationic lipids were used to create ionic interactions between negative charges present on the mucus surface of the GI tract and positive charges on the surface of SLNs. cSLN protected insulin from the enzymatic activity of trypsin and pepsin, improved its stability, and led to sustained release of insulin in the media [30].

Hydroxyapatite (HAP)

Hydroxyapatite (HAP) is an inorganic, biocompatible, and bioactive material and it has been considered as an ideal carrier for drug delivery purposes due to its porosity. Additionally, HAP is an efficient adsorption substance when combined with proteins and nucleic acids [12, 32]. Results showed that PEG-functionalized HAP was effective in the oral insulin delivery system [12]. In the study, insulin was encapsulated into HAP crystal lattice. This carrier led to constant insulin release and regulated blood glucose levels in rats. It was suggested that insulin-loaded HAP might be engulfed bymacrophages and the carrier dissolves in the hybrid of lysosome and endosome. Then, the concentration of PO_4^{3-} and Ca^{2+} would increase which leads to osmotic pressure and disruption of lysosome and endosome hybrids. The high concentration of Ca^{2+} ions also triggers exocytosis of insulin into the extracellular matrix followed by diffusing into the bloodstream and reduction of the blood glucose level [32].

Nanogels

Nanogels are small size hydrogel particles and three-dimensional network nanostructures in the range of 10-1000 nm. Nanogels consist of chemically or physically crosslinked polymer chains. These nanocarriers can improve the release rate of encapsulated drugs and protect them from enzymatic activity. Their nano size permits easy their functionalization and administration through injection [4, 9, 33]. Dextran-crosslinked glucose-responsive nanogels (poly(N isopropy-lacrylamideco-dextran-grafted maleic acid-co-4-(1,6-dioxo-2,5-diaza-7-oxamyl) phenylboronic acid) were synthesized for insulin delivery. These nanogels showed high encapsulation efficiency and the *in vitro* release profiles confirmed a glucose-dependent release of insulin under temperature and physiological pH [33]. Surfactant-free hydroxypropyl methylcellulose (HPMC) nanogels were developed for controlled release of insulin. These HPMC nanogels were pH and temperature sensitive, indicating the encapsulated insulin can diffuse out gradually through the nanogels by adjusting the pH or temperature [9]. A pH-sensitive polyelectrolyte methyl methacrylate (MMA)/itaconic acid (IA) nanogel was designed as a carrier for improving the absorption of insulin. The results of *an in vivo* study showed that these nanogels at a dose could significantly reduce blood glucose level than the control groups [34]. A multiresponsive nanogel was synthesized by N,Ndiethylacrylamide (DEA) and 4-vinylphenylboronic acid (VPBA) as the reaction monomer and loaded with insulin. According to the findings, nanogels loaded with insulin showed high entrapment efficiency and loading efficiency and the release rate of insulin from nanogels could be adjusted by changing the concentration of glucose [35].

Micelle

Polymeric micelles have been found as an appropriate drug delivery system for insulin. A phenylboronic acid- and diol-based block copolymer were synthesized for insulin delivery. The complex micelles increased glucose responsiveness in physiological conditions than simple micelles. Therefore, these nanocarriers can be employed as a promising approach for insulin delivery systems [36]. Glucose and H_2O_2 dual-responsive polymeric micelles were designed for the regulated release of insulin. These polymeric micelles were synthesized by poly (ethylene glycol)-block-poly(amino phenylboronic ester) (PEG-b-PAPBE). Polymeric micelles significantly showed a hypoglycemic effect than free insulin *in vivo*. Furthermore, this polymeric micelle exhibited good biocompatibility without organ damage and parenchymal inflammation [37]. A polymeric complex micelle was prepared by the self-assembly of block copolymer PEG-b-P(Asp-co-AspPBA) and glycopolymer P(Asp-co-AspGA-co-AspNTA) and with a PEG shell. Thispolymeric complex micelle increased insulin loading efficiency, stability, and improved regulated release under physiological conditions [38].

Dendrimer

Dendrimers are highly branched macromolecules that have uniform and controlled size, abundant surface-active properties, and low viscosity in solution. These features have made them ideal nanocarriers for biomedical applications [39, 40]. It has been reported that dendrimers can effectively improve the absorption of drugs in different ways. Caproyl-Modified G2 PAMAM dendrimer (G2-AC) was designed for increasing the pulmonary absorption of insulin. Results showed that G2-AC could efficiently increase the pulmonary absorption of insulin through paracellular and transcellular pathways. Therefore, G2-AC can be applied as a promising absorption enhancer to improve the pulmonary absorption of insulin [41]. Amphiphilic dendrimers based on multiarmed poly (ethylene glycol) modified with benzoboroxole containing hydrophobic blocks were synthesized for insulin delivery. The insulin incorporated in the dendrimers showed strong stability under acidic pH and reduced blood glucose level. Moreover, no hemolysis was observed, which indicated the biocompatibility of dendrimers [42].

Exosomes

Exosomes are membrane nanovesicles with an average size of 30 to 150 nm that are involved in cell-to-cell communication. Their membrane is composed of lipids, proteins, CD9, CD63, CD81, and CD82 [43, 44]. Moreover, exosomes contain nucleic acids, proteins, lipids, fatty acids, tetraspanins, annexins, heat shock proteins (HSPs), Rab (Ras-associated binding protein) GTPases, ESCRT complex (the endosomal sorting complexes required for transport), TSG101 (tumor susceptibility 101) and ALIX (ALG-2 interacting protein X) [43]. Exosomes can be obtained from various cells such as stem cells, immune cells, and cancer cells, and from body fluids; including cerebrospinal fluids and blood [45]. It has been reported that exosomes can represent a valuable delivery system for peptide drugs such as insulin. In a study, cell-derived exosomes were isolated and encapsulated with insulin by electroporation, and their efficiencies were also evaluated. Insulin-loaded exosomes could promote and increase the transport and metabolism of glucose in cells. These results indicated that exosomes can be a valuable nanocarrier for insulin delivery for the treatment of diabctic paticnts [44]. Treatment of mice with exosomes isolated from plasma improved glucose tolerance, insulin sensitivity, and reduced triglycerides plasma level [46].

CONCLUSION

Today, NPs based drug delivery systems are showing a vital role in the pharmaceutical industry. Nanocarriers are employed for loading drugs to limit their adverse effects and improve their bioavailability and efficacy. Recently, insulin delivery systems have been extensively considered for diabetes treatment. NPs can increase the permeability of insulin to the intestinal mucosa *via* opening the tight junctions between intestinal epithelial cells. In the future, nanocarrier based insulin delivery might be replaced with the conventional method such as subcutaneous insulin injection for the treatment of diabetic patients.

REFERENCES

[1]   Volpatti LR, Matranga MA, Cortinas AB, *et al.* Glucose-responsive nanoparticles for rapid and extended self-regulated insulin delivery. ACS Nano 2020; 14(1): 488-97.
        [http://dx.doi.org/10.1021/acsnano.9b06395] [PMID: 31765558]

[2]   Foma MA, Saidu Y, Omoleke SA, Jafali J. Awareness of diabetes mellitus among diabetic patients in the Gambia: A strong case for health education and promotion. BMC Public Health 2013; 13(1): 1124.
        [http://dx.doi.org/10.1186/1471-2458-13-1124] [PMID: 24304618]

[3]   Lin X, Xu Y, Pan X, *et al.* Global, regional, and national burden and trend of diabetes in 195 countries and territories: An analysis from 1990 to 2025. Sci Rep 2020; 10(1): 14790.
        [http://dx.doi.org/10.1038/s41598-020-71908-9] [PMID: 32901098]

[4]   Zhao R, Lu Z, Yang J, Zhang L, Li Y, Zhang X. Drug delivery system in the treatment of diabetes mellitus. Front Bioeng Biotechnol 2020; 8: 880.
        [http://dx.doi.org/10.3389/fbioe.2020.00880] [PMID: 32850735]

[5]   Yazdi JR, Tafaghodi M, Sadri K, *et al.* Folate targeted PEGylated liposomes for the oral delivery of insulin: *In vitro* and *in vivo* studies. Colloids Surf B Biointerfaces 2020; 194: 111203.
        [http://dx.doi.org/10.1016/j.colsurfb.2020.111203] [PMID: 32585538]

[6]   Alai MS, Lin WJ, Pingale SS. Application of polymeric nanoparticles and micelles in insulin oral delivery. Yao Wu Shi Pin Fen Xi 2015; 23(3): 351-8.
        [PMID: 28911691]

[7]   Song M, Wang H, Chen K, *et al.* Oral insulin delivery by carboxymethyl-β-cyclodextrin-grafted chitosan nanoparticles for improving diabetic treatment. Artif Cells Nanomed Biotechnol 2018; 46(sup3): 774-82.

[http://dx.doi.org/10.1080/21691401.2018.1511575] [PMID: 30280608]

[8]   Akinlade AT, Ogbera AO, Fasanmade OA, Olamoyegun MA. Serum C-peptide assay of patients with hyperglycemic emergencies at the Lagos State University Teaching Hospital (LASUTH), Ikeja. Int Arch Med 2014; 7(1): 50.
[http://dx.doi.org/10.1186/1755-7682-7-50] [PMID: 25945127]

[9]   Zhao D, Shi X, Liu T, Lu X, Qiu G, Shea KJ. Synthesis of surfactant-free hydroxypropyl methylcellulose nanogels for controlled release of insulin. Carbohydr Polym 2016; 151: 1006-11.
[http://dx.doi.org/10.1016/j.carbpol.2016.06.055] [PMID: 27474648]

[10]   Sharma G, Sharma AR, Nam JS, Doss GPC, Lee SS, Chakraborty C. Nanoparticle based insulin delivery system: The next generation efficient therapy for Type 1 diabetes. J Nanobiotechnology 2015; 13(1): 74.
[http://dx.doi.org/10.1186/s12951-015-0136-y] [PMID: 26498972]

[11]   Liu J, Hirschberg C, Fanø M, Mu H, Müllertz A. Evaluation of self-emulsifying drug delivery systems for oral insulin delivery using an *in vitro* model simulating the intestinal proteolysis. Eur J Pharm Sci 2020; 147: 105272.
[http://dx.doi.org/10.1016/j.ejps.2020.105272] [PMID: 32084584]

[12]   Zhang Y, Zhang L, Ban Q, Li J, Li CH, Guan YQ. Preparation and characterization of hydroxyapatite nanoparticles carrying insulin and gallic acid for insulin oral delivery. Nanomedicine 2018; 14(2): 353-64.
[http://dx.doi.org/10.1016/j.nano.2017.11.012] [PMID: 29157980]

[13]   He H, Lu Y, Qi J, Zhao W, Dong X, Wu W. Biomimetic thiamine- and niacin-decorated liposomes for enhanced oral delivery of insulin. Acta Pharm Sin B 2018; 8(1): 97-105.
[http://dx.doi.org/10.1016/j.apsb.2017.11.007] [PMID: 29872626]

[14]   Agazzi ML, Herrera SE, Cortez ML, Marmisollé WA, Tagliazucchi M, Azzaroni O. Insulin delivery from glucose-responsive, self-assembled, polyamine nanoparticles: Smart "sense-and-treat" nanocarriers made easy. Chemistry 2020; 26(11): 2456-63.
[http://dx.doi.org/10.1002/chem.201905075] [PMID: 31889346]

[15]   Xiao Y, Tang Z, Wang J, *et al.* Oral insulin delivery platforms: Strategies to address the biological barriers. Angew Chem Int Ed 2020; 59(45): 19787-95.
[http://dx.doi.org/10.1002/anie.202008879] [PMID: 32705745]

[16]   Bashyal S, Seo JE, Keum T, Noh G, Lamichhane S, Lee S. Development, characterization, and *ex vivo* assessment of elastic liposomes for enhancing the buccal delivery of insulin. Pharmaceutics 2021; 13(4): 565.
[http://dx.doi.org/10.3390/pharmaceutics13040565] [PMID: 33923670]

[17]   de Souza Von Zuben E, Eloy JO, Araujo VHS, Gremião MPD, Chorilli M. Insulin-loaded liposomes functionalized with cell-penetrating peptides: Influence on drug release and permeation through porcine nasal mucosa. Colloids Surf A Physicochem Eng Asp 2021; 622: 126624.
[http://dx.doi.org/10.1016/j.colsurfa.2021.126624]

[18]   El Leithy ES, Abdel-Bar HM, Ali RAM. Folate-chitosan nanoparticles triggered insulin cellular uptake and improved *in vivo* hypoglycemic activity. Int J Pharm 2019; 571: 118708.
[http://dx.doi.org/10.1016/j.ijpharm.2019.118708] [PMID: 31593805]

[19]   Dawoud MHS, Yassin GE, Ghorab DM, Morsi NM. Insulin mucoadhesive liposomal gel for wound healing: A formulation with sustained release and extended stability using quality by design approach. AAPS PharmSciTech 2019; 20(4): 158.
[http://dx.doi.org/10.1208/s12249-019-1363-6] [PMID: 30963353]

[20]   Wong CY, Al-Salami H, Dass CR. Formulation and characterisation of insulin-loaded chitosan nanoparticles capable of inducing glucose uptake in skeletal muscle cells *in vitro* J Drug Deliv Sci Technol 2020; 57: 101738.
[http://dx.doi.org/10.1016/j.jddst.2020.101738]

[21]   Dhanasekaran S, Rameshthangam P, Venkatesan S, Singh SK, Vijayan SR. *In vitro* and in silico studies of chitin and chitosan based nanocarriers for curcumin and insulin delivery. J Polym Environ 2018; 26(10): 4095-113.
[http://dx.doi.org/10.1007/s10924-018-1282-8]

[22]   Prudkin-Silva C, Martínez JH, Mazzobre F, *et al.* High molecular weight chitosan based particles for insulin encapsulation obtained *viananospray* technology. Dry Technol 2022; 40(2): 430-45.
[http://dx.doi.org/10.1080/07373937.2020.1806863]

[23]   Agrawal AK, Urimi D, Harde H, Kushwah V, Jain S. Folate appended chitosan nanoparticles augment the stability, bioavailability and efficacy of insulin in diabetic rats following oral administration. RSC Advances 2015; 5(127): 105179-93.
[http://dx.doi.org/10.1039/C5RA19115G]

[24]   Sharma D, Arora S, Singh J. Smart thermosensitive copolymer incorporating chitosan–zinc–insulin electrostatic complexes for controlled delivery of insulin: Effect of chitosan chain length. Int J Polym Mater 2020; 69(16): 1054-68.
[http://dx.doi.org/10.1080/00914037.2019.1655750] [PMID: 33012880]

[25]   Mohammadpour F, Hadizadeh F, Tafaghodi M, *et al.* Preparation,*in vitro* and *in vivo* evaluation of PLGA/Chitosan based nano-complex as a novel insulin delivery formulation. Int J Pharm 2019; 572: 118710.
[http://dx.doi.org/10.1016/j.ijpharm.2019.118710] [PMID: 31629731]

[26]   Wei P, Xu Y, Zhang H, Wang L. Continued sustained insulin-releasing PLGA nanoparticles modified 3D-Printed PCL composite scaffolds for osteochondral repair. Chem Eng J 2021; 422: 130051.
[http://dx.doi.org/10.1016/j.cej.2021.130051]

[27]   Lal S, Perwez A, Rizvi MA, Datta M. Design and development of a biocompatible montmorillonite PLGA nanocomposites to evaluate *in vitro* oral delivery of insulin. Appl Clay Sci 2017; 147: 69-79.
[http://dx.doi.org/10.1016/j.clay.2017.06.031]

[28]   Onugwu AL, Attama AA, Nnamani PO, Onugwu SO, Onuigbo EB, Khutoryanskiy VV. Development and optimization of solid lipid nanoparticles coated with chitosan and poly(2-ethyl-2-oxazoline) for ocular drug delivery of ciprofloxacin. J Drug Deliv Sci Technol 2022; 74: 103527.
[http://dx.doi.org/10.1016/j.jddst.2022.103527]

[29]   Agrawal YO, Husain M, Patil KD, *et al.* Verapamil hydrochloride loaded solid lipid nanoparticles: Preparation, optimization, characterisation, and assessment of cardioprotective effect in experimental model of myocardial infarcted rats. Biomed Pharmacother 2022; 154: 113429.
[http://dx.doi.org/10.1016/j.biopha.2022.113429] [PMID: 36007280]

[30]   Hecq J, Amighi K, Goole J. Development and evaluation of insulin-loaded cationic solid lipid nanoparticles for oral delivery. J Drug Deliv Sci Technol 2016; 36: 192-200.
[http://dx.doi.org/10.1016/j.jddst.2016.10.012]

[31]   He H, Wang P, Cai C, Yang R, Tang X. VB12-coated Gel-Core-SLN containing insulin: Another way to improve oral absorption. Int J Pharm 2015; 493(1-2): 451-9.
[http://dx.doi.org/10.1016/j.ijpharm.2015.08.004] [PMID: 26253378]

[32]   Shyong YJ, Tsai CC, Lin RF, *et al.* Insulin-loaded hydroxyapatite combined with macrophage activity to deliver insulin for diabetes mellitus. J Mater Chem B Mater Biol Med 2015; 3(11): 2331-40.
[http://dx.doi.org/10.1039/C4TB01639D] [PMID: 32262063]

[33]   Elshaarani T, Yu H, Wang L, *et al.* Dextran-crosslinked glucose responsive nanogels with a self-regulated insulin release at physiological conditions. Eur Polym J 2020; 125: 109505.
[http://dx.doi.org/10.1016/j.eurpolymj.2020.109505]

[34]   Mudassir J, Darwis Y, Muhamad S, Ali Khan A. Self-assembled insulin and nanogels polyelectrolyte complex (Ins/NGs-PEC) for oral insulin delivery: Characterization, lyophilization and *in-vivo* evaluation. Int J Nanomedicine 2019; 14: 4895-909.

[http://dx.doi.org/10.2147/IJN.S199507] [PMID: 31456636]

[35]   Yuan S, Li X, Shi X, Lu X. Preparation of multiresponsive nanogels and their controlled release properties. Colloid Polym Sci 2019; 297(4): 613-21.
[http://dx.doi.org/10.1007/s00396-019-04481-x]

[36]   Gaballa H, Theato P. Glucose-responsive polymeric micelles via boronic acid–diol complexation for insulin delivery at neutral pH. Biomacromolecules 2019; 20(2): 871-81.
[http://dx.doi.org/10.1021/acs.biomac.8b01508] [PMID: 30608155]

[37]   Liu X, Li C, Lv J, *et al.* Glucose and H2O2 dual-responsive polymeric micelles for the self-regulated release of insulin. ACS Appl Bio Mater 2020; 3(3): 1598-606.
[http://dx.doi.org/10.1021/acsabm.9b01185] [PMID: 35021650]

[38]   Li C, Huang F, Liu Y, *et al.* Nitrilotriacetic acid-functionalized glucose-responsive complex micelles for the efficient encapsulation and self-regulated release of insulin. Langmuir 2018; 34(40): 12116-25.
[http://dx.doi.org/10.1021/acs.langmuir.8b02574] [PMID: 30212220]

[39]   Nowacka O, Milowska K, Belica-Pacha S, *et al.* Generation-dependent effect of PAMAM dendrimers on human insulin fibrillation and thermal stability. Int J Biol Macromol 2016; 82: 54-60.
[http://dx.doi.org/10.1016/j.ijbiomac.2015.10.029] [PMID: 26598047]

[40]   Fernandes T, Martins NCT, Daniel-da-Silva AL, Trindade T. Dendrimer-based magneto-plasmonic nanosorbents for water quality monitoring using surface-enhanced Raman spectroscopy. Spectrochim Acta A Mol Biomol Spectrosc 2022; 283: 121730.
[http://dx.doi.org/10.1016/j.saa.2022.121730] [PMID: 35988470]

[41]   Yan C, Gu J, Lv Y, *et al.* Caproyl-modified G2 PAMAM dendrimer (G2-AC) Nanocomplexes increases the pulmonary absorption of insulin. AAPS PharmSciTech 2019; 20(7): 298.
[http://dx.doi.org/10.1208/s12249-019-1505-x] [PMID: 31456109]

[42]   Zeng Z, Qi D, Yang L, *et al.* Stimuli-responsive self-assembled dendrimers for oral protein delivery. J Control Release 2019; 315: 206-13.
[http://dx.doi.org/10.1016/j.jconrel.2019.10.049] [PMID: 31672623]

[43]   Ashrafizadeh M, Kumar A, Aref AR, Zarrabi A, Mostafavi E. Exosomes as promising nanostructures in diabetes mellitus: From insulin sensitivity to ameliorating diabetic complications. Int J Nanomedicine 2022; 17: 1229-53.
[http://dx.doi.org/10.2147/IJN.S350250] [PMID: 35340823]

[44]   Rodríguez-Morales B, Antunes-Ricardo M, González-Valdez J. Exosome-mediated insulin delivery for the potential treatment of diabetes mellitus. Pharmaceutics 2021; 13(11): 1870.
[http://dx.doi.org/10.3390/pharmaceutics13111870] [PMID: 34834285]

[45]   Lee J, Lee JH, Chakraborty K, Hwang J, Lee YK. Exosome-based drug delivery systems and their therapeutic applications. RSC Advances 2022; 12(29): 18475-92.
[http://dx.doi.org/10.1039/D2RA02351B] [PMID: 35799926]

[46]   Castaño C, Mirasierra M, Vallejo M, Novials A, Párrizas M. Delivery of muscle-derived exosomal miRNAs induced by HIIT improves insulin sensitivity through down-regulation of hepatic FoxO1 in mice. Proc Natl Acad Sci USA 2020; 117(48): 30335-43.
[http://dx.doi.org/10.1073/pnas.2016112117] [PMID: 33199621]

CHAPTER 4

Nanoscience for Drug Delivery in Diabetes

N. Vishal Gupta[1,*], M. Sharadha[1], K. Trideva Sastri[1], Souvik Chakraborty[1], Hitesh Kumar[1], Vikas Jain[1] and **Surajit Dey[2]**

[1] *Department of Pharmaceutics, JSS College of Pharmacy, JSS Academy of Higher Education & Research, Mysuru, Karnataka, India*

[2] *Roseman University of Health Sciences, College of Pharmacy, Henderson, Nevada, USA*

Abstract: Current conventional diabetes mellitus (DM) therapies are inadequate and have poor patient compliance. Subsequently, it is necessary to explore nanomedicine in managing diabetes. In recent years, several nanocarrier systems have been proven effective in various aspects of diabetes treatment, increasing drug stability, overcoming different biological barriers, and in enhancing bioavailability. Nanomedicine can potentially improve the therapeutic effect of drug substances to gain the patient's belief and impart a greater level of acceptability. In the present scientific spectrum, nanomedicines promise to provide sustained and targeted delivery with potential physical stability for a prolonged period, rendering a safe and effective therapy for diabetes. This chapter comprehensively elaborates on trends in the drug delivery system in treating diabetes for improved delivery of different classes of antidiabetic agents compared to contemporary therapies.

Keywords: Diabetes, Drug delivery, Inorganic nanocarriers, Nanomedicine, Organic nanocarriers, Targeting.

NANOMEDICINE IN DIABETES: NEED-BASED APPROACH

Diabetes mellitus constitutes clusters of metabolic disorders associated with higher blood glucose levels propelled by insulin resistance or deficiency [1]. Every year a substantial number of individuals are affected by this disorder. When the islet β-cell in the pancreas is significantly affected mostly by autoimmune destruction, it leads to Type–1 diabetes [2]. In other circumstances, insulin fails to trigger any response, and insulin resistance contributes to hyperglycemia [3]. Constant monitoring of glucose levels is significant to avoid any downstream impediment in the patients. Studies indicated that sustained hyperglycemia conditions would lead to macrovascular or microvascular complications [4].

* **Corresponding author N. Vishal Gupta:** Department of Pharmaceutics, JSS College of Pharmacy, JSS Academy of Higher Education & Research, Mysuru, Karnataka, India; E-mail: vkguptajss@gmail.com

Ali Rastegari (Ed.)

Traditionally diabetes patient is rigorously monitored for their glucose levels and subsequently administered insulin [5]. Patients are exhaustive with tedious and painful procedures leading to erratic glucose monitoring and meagre adherence. These patient factors precedent to irregular doses ultimately result in seizures and altered glucose levels [6]. There have been tremendous efforts made by researchers to develop continuous glucose monitors along with insulin pumps to tackle these patient difficulties. However, it is necessary to enhance these types of equipment for better management of diabetes. Over the years, there has been exponential growth in nanoscience that deliberated promising results for managing diabetes conditions [7]. Nanotechnology has delivered many breakthroughs explicitly in the medical field by enabling researchers to develop proficient nanosystems for delivering potential therapeutic molecules with enhanced benefits [8]. The principles of nanotechnology are exercised to design nanomedicines as nanotherapeutics; these systems enable the loading of the therapeutic moiety, subsequently enhancing its physiochemical properties, and achieving enhanced therapeutic benefits with precise targeting [9]. Nanotechnology has played a vital role in developing new-age glucose-monitoring devices. Fascinatingly, nanotechnology has potentiated the efforts of scientists in developing numerous delivery systems for improving insulin molecules and other antidiabetic molecules in the systemic circulation, surpassing the usual harsh metabolic pathway that deliberately reduces the efficacy of these molecules; thus, these nanosystems offer a better approach than conventional methods to deliver anti-diabetic molecules [10]. Diabetes is a very peculiar disease, especially type-2, *i.e.*, severely affected by insulin resistance/deficiency. Interestingly, it was discovered that in a subcategory of type – 2, a significant number of patients experience varied blood–glucose levels due to obesity; these effects are independent of insulin. A certain number of patients suffer from insulin deficiency and other sections from insulin resistance [11]. These diabetic conditions have grabbed the attention of scientists, and it is believed that nanomedicine could play a potential role in managing these categories. In recent times, nanotechnological approaches have yielded new–age delivery systems capable of enhancing anti-diabetic molecules' potential [12]. Studies supported the vital role of various nano-formulations specially designed with novel smart polymers that successfully shield the drug molecules from harsh metabolic pathways; subsequently, these systems were instrumental in achieving a controlled release pattern of the loaded molecules, thus facilitating the maintained levels of insulin in patients [13]. The various transport mechanisms available for drug delivery of nanocarriers for the management of diabetes are illustrated in Fig. (**1**). Furthermore, constant monitoring of glucose levels is essential for diabetic patients. A more accurate, highly sensitive, robust nanosensors could be deployed along with other nanomaterials in glucose monitoring devices which would drastically improve the patient's life [14].

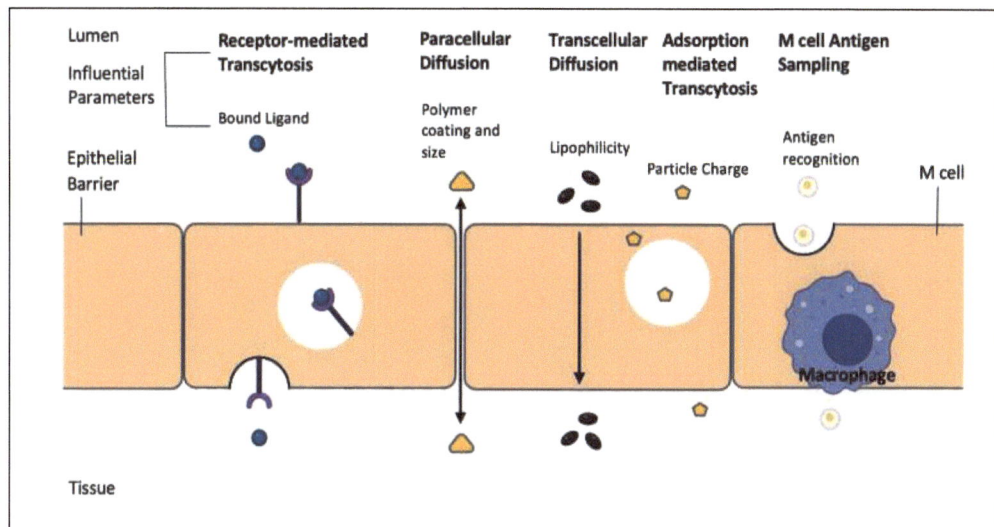

Fig. (1). Various transport mechanisms are available for drug delivery of nanocarriers for the management of diabetes.

Nanomedicine in the Management of Diabetes

Numerous types of nanomedicines have been studied as a drug delivery system for diabetes management as mentioned in Fig. (**2**).

ORGANIC MATERIAL-BASED NANOMEDICINES

Organic nanomaterials are nanocarriers assembled smartly from organic compounds and have drawn significant attention, notably for drug delivery in developing organic frameworks used in biomedical and pharmaceutical nanotechnology. Solid evidence of organic nanocarriers was investigated with lipid-based, natural, and synthetic polymeric nanocarriers.

Lipid-based Nanocarriers

Lipid-based nanocarriers are widely explored as carriers of drugs owing to their remarkable advantages due to their less toxicity, high loading efficiency, good stability, good protectivity, controlled and sustained release, affordable scale-up manufacturing, and targeted site-specific delivery through oral, topical, dermal, parenteral and pulmonary routes. The word lipid-based nanocarriers include liposomes, Solid Lipid Nanoparticles (SLNs), Nanostructured Lipid Carriers

(NLCs), Nano-emulsions, Microemulsions, Nano-capsules, Self-nano-and micro-emulsifying drug delivery systems [15].

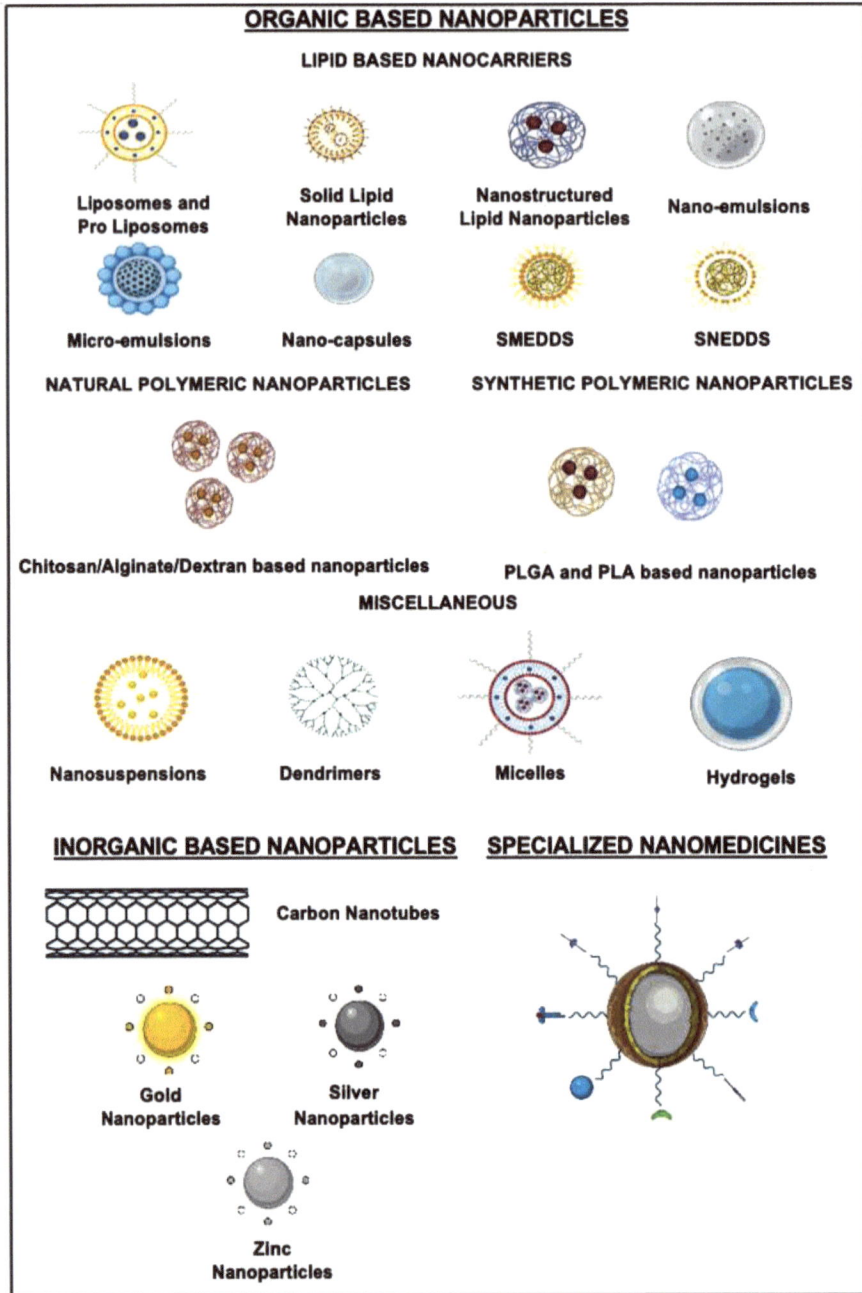

Fig. (2). Various nanocarriers suitable for managing diabetes.

Liposomes and Pro-liposomes

Liposomes are sphere-shaped vesicles comprising one or two phospholipid bilayers surrounded by an aqueous core. They are a powerful drug delivery system for many drugs due to their biodegradable, non-toxic properties with better therapeutic efficacy. They are prepared by many techniques, but the conventional methods used are thin-film hydration, solvent injection, reverse-phase evaporation, and detergent removal method. Hence, liposomes have achieved extensive attention for several drugs as a drug delivery system [16].

Orally administered insulin in treating diabetes mellitus poses a harsh gastrointestinal condition. Sarhadi *et al.* have developed vitamin B12-targeted PEGylated liposomes to deliver insulin *via* the oral route to overcome this problem. Firstly, B12 was conjugated to PEG, then liposomes were formulated by thin-film and extrusion method followed by linking with conjugated B12. The conjugated liposome was characterized and evaluated for particle size parameters, encapsulation parameters, and *in vitro* drug release studies in simulated gastric and intestinal fluid. Stability studies showed that functionalized liposomes were more stable than non-functionalized liposomes in simulated fluids. The *in vitro* cellular uptake studies exhibited significantly enhanced uptake of conjugated liposomes than non-functionalized liposomes in Caco-2 cells with no toxicity. *In vivo* studies with BALB/c mice showed elevated insulin levels in the intestine and liver. Compared to other formulations, B12-targeted PEGylated liposomes showed the highest bioavailability in diabetic rats. Altogether, these research investigations propose that conjugated liposome is an effective formulation to deliver insulin orally [17]. The bioavailability of the poorly soluble orally administered drug is enhanced by loading the drug in liposomes. Wang *et al.* have developed liquiritin (LT) loaded liposomes to enhance bioavailability and hypoglycaemic effects. The *in vitro* drug release was superior in LT-Liposomes than LT suspension due to the increase in solubility, smaller particle size, and larger surface area of LT in the loaded liposomes. The pharmacokinetic estimates exhibited that LT-Liposome has 8.8 times more relative oral bioavailability than LT suspension. In the streptozotocin-induced mouse model, LT-Liposome showed improved hypoglycaemic effect and repair of organs related to diabetes, thereby confirming the antioxidant activity. Besides, biochemical indicators related to diabetes and histopathological changes confirmed further effectiveness of LT-Liposome. These results proved that LT-Liposome had improved LT's solubility, bioavailability, and hypoglycaemic activity *via* the oral route [18].

Pro-liposomes (PL) are an alternative to liposomes to overcome mainly the storage and stability issues associated with liposomes. PLS are dry and free-flowing granular products that form liposomal dispersion on contact or hydration with biological fluids or water. They are composed of phospholipids and water-soluble porous powder. They are prepared by film deposition, spray drying, and supercritical antisolvent, and fluidized bed methods. They are the most preferred drug delivery system to conventional liposomes and vesicular systems because the dry powder form is easy to handle, transfer, distribute, measure, store, and enhance solubility and stability [19]. Glibenclamide (GLB) is a poorly water-soluble drug with a short half-life that shows low bioavailability and needs repetitive dosing. The topical dosage of GLB was reported to show low permeation. It is an ideal candidate to be loaded in pro-liposomes for oral delivery to attain safety, effectiveness, stability, and dissolution. Therefore, Khan *et al.* have formulated glibenclamide-loaded pro-liposomes to increase permeability and *in vitro* release. The permeation studies of GLB-PL showed a 2.27-fold increase in permeability to GLB-liposomal gel, and the *in vitro* release showed 90% release in 12 hrs. The safety of PL was confirmed by acute oral toxicity in Wistar rats. It can be concluded that both GLB-PL and GLB-liposomal gels have a better ability to regulate diabetes mellitus *via* topical and oral routes [20].

Solid Lipid Nanoparticles (SLNs)

SLNs are nanocarriers comprised of lipid cores enclosed by a monolayer of surfactants that stabilize them to form colloidal dispersions at the nanoscale range. Lipids that produce SLNs are fatty acids, triglycerides, phospholipids, or waxes; lecithin, poloxamers, bile salt derivatives, and polysorbates as surfactants. Besides, mucoadhesion properties can be achieved by surface modification of SLNs. The essential peculiarity of SLNs is that they remain solid at ambient and body temperature. They are prepared by numerous techniques, which include high-pressure homogenization, solvent injection, solvent emulsification/ evaporation, and micro-emulsion techniques. In the early 2000s, SLNs gained their attention due to various advantages such as biocompatibility, biode-gradability, enhanced solubility, bioavailability, stability, and ability to encapsulate lipophilic and hydrophilic drugs, and sustained drug release. Thus, SLNs' use has been reported to deliver antidiabetic drugs effectively. For example, Soha *et al.* have formulated Repaglinide (REP)-loaded SLNs integrated *in-situ* gel to enhance the pharmacotherapy of diabetes. The prepared *in situ* gel showed outstanding drug encapsulation efficiency with 83% of sustained drug release. The REP-SLNs proved safe in diabetic rats and showed 1.2-fold greater hypoglycaemic activity through nasal administration than the oral route. Conclusively, maximum therapeutic efficacy was achieved through nasal delivery of REP-SLNs *in situ* gel, reducing the dosing frequency for managing diabetes mellitus [21]. In another study, Noha *et al.* developed Valsartan (VAL) (antihypertensive drug) loaded SLNs integrated *in situ* gel using the homogenization method in managing diabetic foot ulcers.

VAL-SLNs exhibited healing properties through the pathways of COX-2, NFκB, NO, TGF-β, MMPs, and VEGF after 12 days of treatment *via* topical application in diabetes-induced neonatal Sprague Dawley rats. The research was also supported by histological and histomorphometric examinations, which decreased collagen [22].

Nanostructured Lipid Carriers (NLCs)

NLCs are nanocarriers consisting of biocompatible solid or liquid lipid matrix, surfactants, and co-surfactants. It is fabricated by numerous techniques based on high energy (high-pressure homogenization and high shear homogenization), low energy (micro-emulsion, double emulsion, phase inversion, and membrane contractor), and very low energy methods (Emulsification solvent diffusion and evaporation, solvent injection). NLCs are superior to SLNs because of their high entrapment and loading capacity, control release pattern of the drug, long shelf life, prevented enzyme degradation, and sensory masking. Its unique characteristics have extended its use in cosmetic and pharmaceutical applications [23]. NLCs have enhanced prolonged circulation and pharmacodynamics profiles of metformin after oral administration to alloxan-induced diabetic rats. Thus, pharmacokinetic estimates such as AUC, Cmax, and T_{max} of PEGylated NLC prolonged the circulation of metformin in the blood and could be a potential aid in diabetes management [24].

Similarly, the absorption and efficacy of silymarin (SLM) were enhanced by loading into NLCs in type II diabetic patients. NLCs were characterized physically and chemically and found to have 92% entrapment efficiency. *In vitro,* cell culture studies with PAMPA and Caco-2 cells exhibited an increase in the permeation of SLM. The cellular uptake studies showed active processes in NLC's internalization. *In vivo* studies on STZ-induced diabetic mice significantly downregulated blood glucose and triglyceride levels than SLM-free drugs. The authors concluded that the formulation exhibited a hyperalgesia effect on a diabetes-induced model [25].

Nano-emulsions

Nano-emulsions (NEs) are a versatile, kinetically stable colloidal system of two non-miscible liquids with 100 nm droplet size. It mainly comprises of lipid phase, an aqueous phase, and an emulsifying agent. It is fabricated by high-pressure homogenization, ultrasonic emulsification, micro-fluidization, spontaneous emulsification, and phase inversion temperature. It increases bioavailability and physical stability to non-toxic, non-irritant and requires less energy. Hence, it is considered safe and effective as a biomedical nanocarrier [26]. The release of glucose slowly after digestion was achieved by double-layer NEs. Razie *et al.*

devised a double-layer NE composed of 0.12% zein and sodium alginate. The glucose release from single and double-layered NEs was evaluated. Double-layered NEs exhibited 240 min of controlled release of glucose under GI conditions, which serve as a potential candidate for patients with hypoglycaemia [27]. The solubility of polyphenols is also enhanced by loading in nano-emulsion. Quercetin NEs are one of them which were effective in diabetic-induced cardiac toxicity. The research findings demonstrated that treatment with Quercetin NEs elevated enzymes and cardiac index in the STZ-induced diabetic group than in the control group. Hence, the results proved that polyphenols NEs effectively manage diabetic-induced heart failure [28].

Microemulsions

Micro-emulsions (MEs) are a class of clear, thermodynamically stable dispersions prepared upon spontaneous mixing of water, oil, and surfactants/co-surfactants. The fabrication methods involve the Winsor phase diagram, pseudo-ternary phase diagram, and fish-like phase diagram. The advantages of these systems include high stability, no requirement for external energy supply, long shelf life, the ability to solubilize organic matter and oil, and a high degree of solubilization [29]. Therefore, MEs offer a broad range of therapeutic applications in targeted and sustained drug delivery.

As mentioned earlier, oral deliveries of insulin are associated with some barriers. To overcome this, researchers developed insulin-loaded microemulsions (MEs) incorporating snail mucin for the oral route. The sustained release of insulin was achieved using a novel strategy by loading mucin in insulin into the internal core of the formulated W/O microemulsion—the developed MEs enhanced stability and drug absorption in the GIT with 70% encapsulation efficiency. Insulin-MEs efficiently reduced blood sugar levels in diabetic rats for up to 8 hrs through the oral route. Hence, it was concluded as a good dosage form for the delivery of oral protein [30].

Similarly, in another study, Kaur *et al.* developed W/O/W Insulin MEs containing piperine and albumin as permeation enhancers and stabilizers for oral insulin administration. The *in vitro* cellular uptake studies exhibited higher uptake than free insulin. Besides, permeability and *ex vivo* intestinal studies also showed a four-fold enhancement in permeation with free insulin. Likewise, pharmacokinetic and pharmacodynamic studies exhibited a 1.6 fold higher AUC and 3.2 fold greater hypoglycaemic effect than free insulin. Conclusively, this strategy can deliver macromolecules orally [31].

Nano-capsules

Lipid nano-capsules (LNCs) are smart lipid nanocarriers, structurally a hybrid of liposomes and polymeric nanoparticles in the nanorange (20-100 nm). They comprise a drug-containing oil core, an aqueous phase, and a lipophilic surfactant surrounded by a polymeric shell. The fabrication of nano-capsules includes the formulation of nano-emulsion followed by polymeric shell formation. The main advantages include biocompatibility, biodegradability, and higher entrapment efficiency than conventional nanoparticles. Hence, the smart generation nanocarriers can be effectively employed for drug delivery, theranostics, and diagnostics [32].

More recently, El-Hussien *et al.* have formulated PLGA polymeric nano-capsules loaded with chrysin to improve oral delivery. The nanocapsule particle-size 176 nm had a negative charge with 87% entrapment efficiency. *In vitro* release studies showed a controlled release of chrysin for 24 hrs from nano-capsules than chrysin suspension and were stable for three months. After oral administration of a dose of 20 mg/kg to diabetes-induced rats, the authors observed a prolonged anti-hyperglycaemic effect compared with the suspension-treated group. Besides, they also reported a marked anti-hyperlipidaemic effect for 28 days [33]. Another research reported the formulation of calcium-alginate nano-capsules coated with chitosan loaded with liraglutide for sustained oral release. The prepared nano-capsules showed 92.5% entrapment efficiency and stability 0ver 60 days. *In vitro* studies in simulated GI conditions exhibited 59% release after 6 hrs. The novel nano-capsules loaded with liraglutide hold great promise in diabetes management [34].

Self-nano-and Micro-emulsifying Drug Delivery Systems (SMEDDS and SNEDDS)

SEDDS was first coined by Panton *et al.* (1985) as emulsion pre-concentrates and has grown massively in the last 15 years. SNEDDS are the anhydrous form of nano-emulsions composed of drugs, oil, and surfactants/co-surfactants. They exhibit long-term stability as they do not contain water and are superior to conventional nano-emulsions. SNEDDS are incorporated in soft/hard/HPMC capsules with low volume to enhance palatability and patient compliance. SNEDDS form nano-emulsions in contact with aqueous media with mild stirring.

In the same way, SMEDDS form micro-emulsions in the presence of an aqueous environment. The conventional SNEDDS and SMEDDS pose long-term incompatibility with capsule shell or precipitation of the drug at a low temperature. So, solid SNEDDS and SMEDDS are preferred for long-term storage and stability [35].

In this line, Bravo-Alfaro *et al.* have developed oral insulin SNEDDS modified with phosphatidylcholine to enhance the bioavailability of conventional insulin. The results showed a particle size of 20-44 nm with 35.7% bioavailability. The authors also observed a reduced blood glucose level of 36% with 1.8% bioavailability in diabetes-induced rats after administration of modified SNEDDS for 4 hrs *via* oral route rather than subcutaneous application [36]. In another example, Singh *et al.* have devised a lipid-containing solid SMEDDS loaded with Canagliflozin (CGF) to manage type-2 DM. The CLSM showed the permeability of CGF into the jejunum segment of the rat. Besides, the CGF-SMEDDS was compared with pure CGF and marketed products and exhibited a 3.9, and 2.5 fold enhancement in C_{max} and a 2.2 and 1.5 fold increase in AUC_{0-24} hrs. In a nutshell, CGD-SMEDDS are well suited for antidiabetic and synergistic activity with CGF [37].

Natural Polymeric Nanoparticles

Natural polymers are derived from plants, animals, fungi, and bacteria, mainly categorized as polysaccharides (chitosan, alginate, dextran, pectin, xanthan gum, cyclodextrins) and protein-based polymers (silk, elastin, gelatine, collagen, albumin, lectin). Natural polymers have more advantages than synthetic ones because they are obtained from a natural source, are inexpensive, and can be modified chemically. Polymer-based polymers have considerable attention due to their biocompatible, biodegradable, mucoadhesive properties and high loading efficiency [38]. In this section, frequently used natural polymers for drug delivery in diabetes management are discussed below.

Chitosan-Based Nanoparticles

Chitosan (CS) is a biocompatible and biodegradable polymer that is considered safe and low toxic for dietary use in humans. As polymeric nanoparticles, it is used for drug delivery *via* several routes of administration. The chemical functional group of chitosan is modified physically and chemically to target specific goals for potential applications. The CS and CS derivatives NPs contain a positive surface charge and good mucoadhesive properties to release the drug in a sustained manner. The drug release from CS-NPs from the polymeric matrix follows diffusion, swelling, and erosion. Due to its versatility, CS can effectively deliver the drug to the target site *via* GIT [39].

More recently, a novel cross-linked carboxymethyl CS-NPs was synthesized and loaded with metformin hydrochloride (MEH) by employing microfluidics (MF) and $CaCl_2$ as a cross-linker. The results exhibited an average size of 77 nm through FE-SEM and DLS studies with 90% encapsulation efficiency. Authors also reported that the formulation treated in diabetes-induced rats enhanced body weight (7.94%) by decreasing blood glucose levels (43.5%) than the free drug. Additionally, histological studies showed the pancreatic islet size and β cell intensity to be 2.32 μm2, and 64 cells/islet in the formulation-treated group than the free drug. All these data proved that carboxymethyl CS-NPs containing MEH were effective in diabetic therapy [40]. In another study, CS-NPs were developed and loaded with polydatin (PD) oral formulation. The *in vitro* studies showed 20% prolonged release after 12 hrs, and their safety was confirmed in Vero cell lines. Moreover, antidiabetic activity was highly significant in diabetic-induced rats treated with PD-CS-NPs and proved a promising non-toxic nanocarrier [41].

Alginate-Based Nanoparticles

Alginate is a natural anionic polymer extracted from brown marine algae. Due to their biodegradability, biocompatibility, chelating, mucoadhesive properties, and low cost, they are used widely in the food and pharmaceutical industry. Besides, hydro-gelling behaviour has broadened its use in biomedical applications by swelling and releasing the encapsulated drug.

The preparation method of Alginate nanoparticles (Al-NPs) includes emulsification/gelation, solvent displacement and evaporation, complexation, spray drying, electrospray, and electrospinning methods. Hence, Al-NP is a fast-developing field mainly of gelling nature and thermostability [42].

In one of the studies, vildagliptin (VG) was loaded into an acrylamide grafted psyllium and alginate core-shell nanoparticles. The cytotoxicity studies on the A549 (human cell adenocarcinoma) cell line revealed no toxicity. Glucose uptake studies on TNF-α-induced insulin in the L6 (Mouse muscle) cell line showed a highly significant increase in glucose uptake after treatment with the formulation. The drug release from the formulation showed 98% drug release at pH 7.4 with the Higuchi release pattern. Hence, the study provides an environment and eco-friendly approach to VG-alginate NPs for anti-diabetes management [43].

Similarly, vitexin was loaded in spray-dried alginate NPs compacted with PEG and stearic acid. The release of vitexin from PEG-compacted NPs was reduced in the stomach than Stearic acid compacted NPs. Altogether, compaction with PEG 10,000 resulted in PEG-NPs interaction that nullified the initial release of vitexin. Therefore, the dissolution of PEG in the intestinal phase resulted in the breakdown of particles and the release of vitexin. Hence, the study showed oral intestinal-specific release of vitexin by lowering blood glucose levels and enhancing intestinal vitexin *in vivo* [44].

Dextran-Based Nanoparticles

Dextran (Dex) NPs are grabbing the attention of technologists and scientists for their bio-materialistic applications in the medical and pharmaceutical fields. Dextran, an α-glucan, natural polysaccharide, and bacterial exopolysaccharide, shows high biodegradability, biocompatibility, good solubility, and non-immunogenicity. Derivatives of dextran have been developed *via* chemical modification due to the presence of larger reactive hydroxyl groups on the dextran backbone. The fabrication methods for dextran-based drug delivery systems include self-assembly, emulsification, co-precipitation, and other strategies. The physicochemical properties, release patterns, and therapeutic effects have been justified in biomedical applications, especially for treatment of diabetes [45].

Besides, the authors reported that dextran surface-modified zein nanoparticles loaded with insulin and cholic acid increased oral insulin absorption in the intestine and liver. The zein in NPs acted as cement to surround insulin, casein, and cholic acid by a hydrophobic bond. Casein was conjugated to hydrophilic dextran by Maillard reaction and was placed on the NP surface. The oral bioavailability of NPs in Type I diabetic-induced mice was 12.5-20% and showed a hypoglycaemic effect of insulin [46]. Another study revealed that insulin-dextran NPs immobilized with glucose oxidase on acryloyl crosslinked dextran dialdehyde (ACDD) NPs was a promising strategy to overcome disadvantages associated with subcutaneous insulin therapy [47].

Synthetic Polymeric Nanoparticles

Synthetic polymers have gained significant attention for the delivery of proteins and peptides. Generally used synthetic polymers are polyesters (PLGA), polyethers (PEG), poloxamers, and recombinant protein-based polymers. They enhance pharmacokinetics and time of circulation of encapsulated therapeutics. As a drug carrier, these polymers exhibit passive function [48]. In this section, frequently used synthetic polymers for drug delivery in diabetes management are discussed below.

PLGA (Poly Lactic-co-Glycolic Acid) Based Nanoparticles

PLGA is a copolymer of PLA and PGA in an equal ratio. It belongs to biodegradable polymers approved by the FDA and has excellent biodegradability and biocompatibility. It follows a non-linear pattern with a dose-dependent profile. The co-polymerization of PLGA with other polymers may enhance the entrapment efficiency, drug release, and stability of entrapped macromolecules. Considerable research has investigated PLGA-based NPs to deliver therapeutics, including macromolecules, through different routes.

In this line, Ren *et al.* have developed PLGA-NPs loaded with exenatide (EXE) to treat diabetes. The double emulsion solvent diffusion technique showed nanosize and positive surface charge with gastrointestinal stability. *In vitro,* cell line studies exhibited effective endocytosis mediated by β-glucan receptor and transport of EXE-PLGA-NPs by the macrophagic absorptive pathway. *In vivo,* pharmacokinetic studies after oral administration of EXE-PLGA-NPs showed a hypoglycaemic effect by enhancing pharmacological availability (13.7%). Hence, the PLGA system showed high therapeutic efficacy in delivering peptide drugs [49]. The solubility and bioavailability of bioflavonoids were enhanced by encapsulating them in PLGA-NPs. In another study, naringenin (N), a bioflavonoid, encapsulated in PLGA-NPs, was evaluated for its antidiabetic potential on diabetic-induced rats and compared with free Naringenin. Blood glucose level was evaluated after treating diabetic rats with 10 mg (i.p.) of free Naringenin and N-PLGA-NPs. A significant reduction in blood glucose level, improvement in oxidative stress, and dyslipidaemia were observed after ten days of treatment with N-PLGA-NPs than free Naringenin. All these findings exhibited the good antidiabetic potential of bioflavonoid encapsulated in NPs [50].

PLA Based Nanoparticles

Polylactic acid (PLA), an aliphatic polyester, gained a prominent role in biomedicine because of the presence of ester bonds connected by the monomer units [51]. El-Naggar *et al.* have developed curcumin (CUR)-loaded PLA-PEG-NPs to manage liver inflammation. It was formulated using the nano-emulsification method by loading hydrophobic CUR in hydrophobic co-polymers with cationic surfactants. CUR-PLA-PEG-NPs and free CUr were administered to diabetic rats and assessed for blood parameters such as ALT, AST, NF-κB, GSH, MDA, NO, COX-2, PPAR-γ and TGF-β1. CUR-PLA-PEG-NPs and free CUr have exhibited negative changes in diabetic rats, but CUR-PLA-PEG-NPs showed a superior effect over free CUr. Histopathological studies in liver tissue showed reduced inflammation on treatment with CUR-PLA-PEG-NPs [52].

Miscellaneous

Nanosuspensions

Pharmaceutical nanosuspensions are nanosized dispersions of heterogeneous aqueous solutions containing insoluble drugs mixed with surfactants to be stable. Potential advantages of nanosuspension are that it reduces the cost of therapy and the dose required for the treatment and minimises the fast state fluctuation in the plasma level. It also surpasses the possibility of dose dumping in the patient's body. These advantages have driven towards faster development of nanosuspension technology in treating diabetes in recent years [53].

Specific advantages of nanosuspension technology have been explored by Gaur, which helped deliver antihyperglycemic flavonoids at their site of action for diabetic therapy. The developed nanosuspensions were evaluated against type 2 diabetes in diabetic-induced rats. The kinetic profile and tissue architecture of the organs after oral administration of the nanosuspensions were determined by pharmacokinetic and histopathological parameters. The flavonoid-loaded nanosuspensions showed optimum effects on the disease in physical parameters. In contrast, pharmacokinetic parameters showed an enhancement in the absorption of flavonoids, with a decrease in the metabolism of active flavonoids. In cool conditions, these formulated nanosuspensions were found to be more stable. Hence, it proves that oral administration of flavonoid-loaded nanosuspensions can potentially treat diabetes [54]. Hemalatha and Monisha have developed a nanoprecipitation method for the preparation of nanosuspensions incorporated with repaglinide to lower glucose level with an optimum drug dose in smaller particle size and to test the optimum clinical efficacy *in-vivo* in albino Wistar rats. The results proved the nanosuspensions to be efficient in lowering postprandial blood glucose levels and delivering consistent release of the drug, which is evident from the constant lowering of glucose levels. The prepared nanosuspensions were very potent and clinically efficient compared to the pure drug and drug suspensions [55].

Dendrimers

Dendrimers can be used as a potential drug delivery system for their surface engineering and toxicological profile, with their effective delivery of antidiabetic drugs to cure diabetes. Dendrimers are three-dimensional, hyperbranched macromolecules, which are nanosized, having radially symmetric molecules with a well-defined, homogeneous, and monodisperse structure consisting of tree-like arms or branches. There are various types of dendrimers such as Poly (propylene imine) dendrimers, Poly(amidoamine) (PAMAM) dendrimers, Frechet-type dendrimers, Chiral dendrimers, Liquid crystalline dendrimers, Peptide dendrimers, Multiple antigen peptide dendrimers, Glycodendrimers, Hybrid dendrimers and Polyester dendrimers. Recent research has developed different types of dendrimers in providing antidiabetic delivery to the patient dealing with the disease [56, 57]. PAMAM dendrimers are nano-sized, highly branched polymers of dendrimers that can be used for potential applications in nanomedicine, including cellular delivery of drugs having a low molecular weight and can be used in nucleic acid-based therapeutics. For the development of diabetes-induced vascular dysfunction, the EGFR-ERK1/2-ROCK signalling pathway has been found as a critical pathway [58].

In previous studies, Akhtar *et al.* observed whether PAMAM dendrimers could be beneficial for inhibiting the critical EGFR-ERK1/2-ROCK signalling pathway when administered chronically and preventing the development of diabetes-induced vascular dysfunction. The collective data of the study revealed for the first time that chronic *in vivo* administration of PAMAM dendrimers in an animal model of diabetes inhibited the EGFR-ERK1/2- ROCK signalling pathway—a detrimental pathway known to be critical in the development of diabetes-induced vascular dysfunction [59]. In another study by Zhang and Huang, PAMAM dendrimers have been connected with polysaccharide hyaluronic acid (HA) through the substrate polypeptide (Gly-PLGLAG-Cys) of MMP-2 to produce MMP-2-responsive nanosuspensions containing HA-pep-PAMAM. For controlled release at the site of intractable wounds, astragaloside (ASI), which is insoluble, was encapsulated in these nanosuspensions. The HA-pep-PAMAM-ASI nanosuspensions were successfully developed with the desired nano-size diameter. A stronger expression of MMP-2 was observed in hard-to-heal wounds of diabetic mice by immunohistochemical staining of the skin than in the wounds of normal mice. On the proliferation of BJ and HaCaT cells, a dose-dependent effect of H_2O_2 was observed, and treatment with HA-pep-PAMAM-ASI showed the best antioxidant capacity with MMP-2 pretreatment. GSH levels have been significantly increased by the formulated nanosuspensions, with a reduction of reactive oxygen species (ROS) levels for achieving antioxidant effects. HA-pe--PAMAM-ASI group showed higher expression of all wound-repair-related genes than the ASI group, with a pronounced *in vivo* therapeutic effect. Therefore, these results showed that enzyme-responsive MMP-2-loaded PAMAM nanoparticles could promote wound healing and may be a promising biomaterial for the treatment of diabetes [60].

Micelles

Micelles are vesicles composed of amphiphilic co-polymers, with a size of 10–100 nm, with the capability to self-assemble under aqueous conditions and form a spherical core-shell structure. Micelles are composed of amphiphilic molecules of two completely different regions, with opposite affinities to water substances. Hydrophobic fragments construct the core of the amphiphilic molecules of the micelles. Micellar amphiphilic molecules at low concentrations exist separately in an aqueous medium [61].

For diabetic patients, administration of exogenous insulin is a necessary therapy, which also involves multiple daily subcutaneous injections, which is very painful and sometimes inconvenient for many patients. On the other hand, the development of insulin formulations for long-acting effects has been hampered by short pharmacokinetics and reduced bioactivity due to long-term storage. To solve this issue, the development of a formulation with high efficiency of insulin loading has been carried out by Xin *et al*. by self-assembling an amphiphilic polymer (P_1M_{10}) with insulin. P_1M_{10} mixed insulin micelles showed quick and continuous control of the glucose level in the blood when applied to type 1 diabetic rat model, without causing any risk of hypoglycemia, with a better therapeutic index than commercially available long-acting insulin. In the long term, P_1M_{10} polymer can protect insulin when mixed in solution at room temperature. For this reason, this kind of micelles can be lyophilized and reconstituted without any loss in the bioactivity of insulin, which allows transportation and storage conveniently for the therapeutic proteins under normal conditions, which makes them acceptable for application in diabetic patients [62].

As less work has been carried out on treating diabetes and vascular diabetes complications, the fabrication of a novel hydrogel based on polysaccharide-based micelles has been carried out by Wen *et al*. to produce synergistic therapy. With the help of the single electron transfer living radical polymerization (SET-LRP) method, Zwitterionic dialdehyde starch-based micelles (SB-DAS-VPBA) were synthesized. Sulfobetaine (SB) and hydrophobic 4-vinylphenylboronic acid (VPBA) have been grafted to dialdehyde starch (DAS) as a backbone, which falls under the segment of hydrophobic and hydrophilic substances individually. Insulin and nattokinase were loaded into the micelle hydrogel for giving synergistic therapy. *In vitro* drug delivery and dissolution of blood clot behaviour have been determined. The results suggested that synergistic therapy by micelle hydrogel possesses delivery of glucose-responsive insulin. Thus, this micelle-hydrogel synergistic therapy system can be used as a treatment platform for diabetes and vascular diabetes complications [63].

Hydrogels

Hydrogels have played an essential role in designing biomaterials for their physical properties, controllable degradability, and ease of fabrication. Because hydrogels contain a large amount of water, they can be excellent biocompatible substances to encapsulate hydrophilic drugs, such as insulin and other materials within their matrix structure. The cross-linked network can also be tuned to restrict the penetration of external proteins to protect bioactive therapeutics from degradation by inwardly diffusing enzymes. As such, enormous advances have been made in hydrogel-based delivery of therapeutics [64].

Zou *et al*. have prepared thermosensitive hydroxypropyl chitin (HPCT) hydrogels to investigate their chemical structure, microstructure, rheological properties, and degradation properties through *in vitro* studies. Formulated hydrogels revealed satisfactory biocompatible properties, where *N*-acetylglucosamine (NAG) and carboxymethyl chitosan (CMCS) polymers were used, which have desired capacity to promote cell proliferation. The effects of HPCT/NAG and HPCT/CMCS thermosensitive hydrogels as RIN-m5F cell carriers were most importantly evaluated, which were applied with the help of injection into different areas of the diabetes-induced rat model. Our results demonstrated that both hydrogels loaded RIN-m5F cells could be helpful in the survival of the cells, maintaining the secretion of insulin and reducing blood glucose levels. Overall, the functional thermosensitive hydrogels based on HPCT were adequate as a cell carrier for RIN-m5F cells and can be used as a novel strategy for treating diabetes through cell engineering [65]. Arepaglinide-loaded hydrogel particles of carboxyethyl xanthan gum (CEXG) and carboxymethyl xanthan gum (CMXG) were fabricated by Patel *et al.* to reveal or achieve controlled drug delivery. This study swelled the hydrogel particles to the maximum size with optimum particle charge. Acetylation of hydrogel particles reduced the drug entrapment efficiency, whereas it also extended drug release, obeying anomalous diffusion. DSC and X-ray diffraction analysis have characterised the amorphous dispersion of repaglinide after entrapment. Acetylated hydrogels caused a considerable reduction in blood glucose levels in the preclinical study. Hence, CMXG hydrogel particles could help control diabetes [66].

INORGANIC MATERIAL-BASED NANOMEDICINE

Inorganic nanoparticles are widely used for drug delivery for various diseases. With the simplest reaction process, thermal decomposition, and metal precursors, the synthesis of inorganic nanoparticles is easy with several methods. Inorganic NPs could be effortlessly accumulated and reached in the liver, lungs, kidneys, and pancreas. Many inorganic nanomedicines include carbon nanotubes, mesoporous silica-based nanoparticles, gold nanoparticles, silver nanoparticles, iron oxide nanoparticles, and quantum dots.

Carbon Nanotubes (CNTs)

Carbon nanotubes (CNTs) are tiny cylindrical tubes made entirely of carbon atoms. It may consist of single or multilayered tubes called single-walled- or multiwalled CNTs, with a sizer of 2 to 100 nm [67]. CNTs are insoluble in water and other organic solvents. It can be easily modified on its surface with chemical interactions, which make it hydrophilic and aqueous soluble. Furthermore, the surface of CNTs can be functionalized with ligands or targeting agents such as pachymic acid [68] and polyethyleneimine-candesartan [69] for site-specific targeting. Multiwalled carbon nanotubes, which are electrically conductive, are used to construct the microphysiometer [70]. The traditional detection techniques usually assess insulin levels at intervals by routinely collecting and evaluating tiny samples. By measuring the transfer of electrons created when glucose oxidises insulin molecules, the microphysiometer may continually and indirectly measure insulin levels [71]. Due to their emission at near-infrared wavelengths, which is a wavelength at which the skin is particularly transparent, carbon nanotubes are excellent candidates for fluorescent biosensing. One of the first to employ single-walled nanotubes (SWNTs) for fluorescent glucose sensing was Barone *et al.* [72]. It was discovered that this sensor was completely reversible and responsive to clinically significant glucose concentrations (1-8 mM). To identify combinations that produce precise and reversible changes in fluorescence signals in response to glucose levels, Yum *et al.* explored 30 boronic acids with SWNTs [73]. The fluorescence signal grew as glucose bound to numerous SWNTs modified by boronic acid. By analysing the frequency shift of the fluorescence or the "switch on" response, the 4-cyanophenylboronic and 4-chlorophenylboronic acids were discovered to have the best combination of desirable qualities.

Metallic Nanoparticles (MNPs)

Metallic nanoparticles (MNPs) with the potential to be used in the prevention and treatment of diseases include magnetic, silver, and gold NPs [74]. Because of their extraordinary and adaptable biophysical characteristics that rely on their size and shape, NPs and other nanomaterials are increasingly being employed in biological applications [75]. The improved catalytic activity of the NP, which increases their capacity to change the substrate, is considered connected to the high ratio of electrons that stay on their surface [76]. Due to their increased surface area, smaller NPs have more robust catalytic activity than more significant NPs. Due to these distinct characteristics of NPs, their therapeutic use is anticipated to offer several advantages over traditional treatment techniques that employ medications with numerous adverse effects due to their inadequate and off-target action [77]. Due to their medical uses, the antihyperglycemic potential of several metallic nanoparticles generated from various biological substances has been revealed to be substantial. MNPs appear to be the need of the hour in disease control due to their minimal side effects, toxicity, and efficacy at the nanoscale level [78].

Due to their antibacterial activity and improved wound-healing benefits, metal nanoparticles are frequently employed in treating diabetic wounds. The activity of -amylase and -glucosidase was more dose-dependent than the pure plant extract, suggesting that the silver nanoparticles of Argyreia Nervosa, Punica granatum, and marine alga Colpomenia sinuosa exhibit antidiabetic potential [79, 80]. These encapsulated metabolites NPs boosted their target selectivity, increasing their diabetes control and management efficacy. According to Prabhu *et al.*, the antidiabetic potential of Ag NPs from Pouteria sapota leaves extract affected changes in biochemical parameters, significantly inhibited -amylase as well as non-enzymatic glycosylation of rat haemoglobin, and increased glucose uptake in yeast cells. Rat histopathological parameters showed that the rat liver had a typical architecture with normal hepatocytes, and the rat kidney had an architecture with normal glomerular. Similarly, Abideen and Vijaya Shankar investigated the antidiabetic effect of silver nanoparticles using Gracillaria edulis and Syringodium isoetifolium, two common seaweeds [81].

Zinc's antidiabetic and insulin-like impact has been described in various *in vitro* and *in vivo* investigations; however, the molecular basis underlying zinc's insulin-like effect has not been well investigated. However, according to other research, zinc oxide NPs interact with insulin signalling pathways and controls glucose metabolism [82]. Zinc oxide NPs increase the phosphorylation of the insulin receptor B-subunit, which activates phosphatidylinositol 3-kinase and protein kinase B [83, 84]. Gold nanoparticles (GNPs) have been widely explored due to their superior characteristics in diagnostics, treatments, and molecular nanoprobes. GNPs from Cassia auriculata [85] and Sargassum swartz [86] showed antihyperglycemic activity in diabetes-induced rats. GNPs from C. auriculata effectively controlled the metabolic parameters of diabetic rats and suppressed PTP1B enzymatic activity. Similarly, the antidiabetic effect of GNPs from Gymnema sylvestre [87] and Fritillaria cirrhosa [88] in diabetic rats was observed, and all biochemical parameters of diabetic rats were restored to normal levels.

SPECIALIZED NANOMEDICINES

In recent years, nanotechnology has discovered favourable conditions for creating innovative delivery systems that may improve the efficacy of antidiabetic regimens. Different formulations of smart materials were developed to protect the medicine by encasing it in a nanocarrier system and effectively releasing the drug in a controlled and progressive way. As a result, individuals with SIRD can get antidiabetic medications that increase insulin synthesis or limit insulin release, while patients with insulin shortage can receive insulin nano-formulations [89, 90]. NP-drug delivery systems that use various strategies to target particular cells better include targeting active, passive, and stimuli-responsive [91]. In order to actively target the cell, a specific ligand (antibody, peptide, aptamer, *etc.*) must be attached to the nanoparticle and recognised by receptors on the membranes of specific receptors. The absorption of the nanomaterial into the cell is made more accessible by ligand binding to the receptors [92]. This approach depends on the production of specific biomarkers often found on the cellular membrane. According to the research by Kaasalainen *et al.*, the production circumstances have a critical role in increasing the release of the glucagon-like peptide (GLP-1) from porous silicon (PSi) NPs [93].

Similarly, Araujo *et al*. incorporated the GLP-1 peptide within various nanomaterials made of various polymers such as PSi, Witepsol, E85 lipid, and poly(lactide-co-glycolide) polymer. Moreover, it examined the *in vitro* susceptibility [94]. Moreover, the liposomal system comprised of DSPE-PG8G, DPPC, DPPG and cholesterol incorporated with GLP-1 attributed to its ability to induce serum GLP-1 level secretion up to 70% [95]. Similarly, GLP-1 functionalized GNPs enhance the bio-compatibility and absorptions against Caco-2 cells. Moreover, the GNPs enhanced the insulinotropic activity of incretin [96]. Li *et al*. reported that chitosan-based nanocarrier systems functionalized with L-valine had higher absorption in the small intestine in diabetic conditions [97, 98]. Similarly, the (FcRn)-mediated silicon NPs could improve insulin transport in the intestine [99].

Natural Carrier Systems-based Nanomedicines

Natural carriers are naturally occurring entities inside the body that regulate glucose homeostasis and the delivery of insulin. Nanomedicine and the use of natural carrier systems such as erythrocytes, exosomes, phytosomes, and sphingosomes can be engineered to sheath the therapeutics moieties such as insulin for precise and controlled drug delivery to the target site, thereby lessening the side effects and increasing the treatment efficacy.

Erythrocytes-based Nanocarriers

Erythrocytes (ER)/Red blood cells (RBC) are natural biological potential carriers responsible for the transport of gases to the tissues from the lung [100]. Nanoerythrosomes (NER) are carrier erythrocytes that emerged as smart and novel drug delivery systems. As a nanocarrier, they possess unique properties such as enhanced bioavailability, biocompatibility, prolonged circulation half-life, enhanced pharmacokinetics, high loading capacity, capability to enter the immune system, and less toxicity [101]. One of the assuring treatments for diabetes is glucose-responsive insulin delivery. But the current systems produce short-term effects. To overcome this limitation, Xu *et al*. have designed a biomimetic erythrocyte-glucose-responsive system (ERGS) by pairing with glucose-responsive NPs (GRNPs) to RBCs. The generated NPs showed dual functionality of persistent incidence in circulation and glucose responsiveness. GRNPs are produced by encapsulating insulin *via* ion cross linkages coloaded with catalase and glucose oxidase, a method that provides GRNPs. Concurrently, the GRNPs are paired with RBCs from the immune system to conceal them to obtain long-term blood circulation. In hyperglycemic conditions, the action of glucose oxidase on blood glucose generates gluconic acid that separates GRNPs to release insulin efficiently whereas, in hypoglycemic conditions, only insulin is released at the basic rate. Therefore, GRNPs can continuously and efficiently modulate insulin release to sustain a normal range of blood glucose levels [102].

Exosomes-based Nanocarriers

Exosomes (EXOs) are extracellular nano-sized micro-vesicles secreted naturally by cells. They exhibit a major role in cell-to-cell contact and deliver therapeutic materials from the original cells. EXOs in diabetes are generally sourced from urine and serum. They are moderately stable and permit prolonged storage thereby offering structural integrity to bioactive molecules. They also enhance the activities and levels of GLUT1 and GLUT4 to aid the uptake of glucose and improve its metabolism. Moreover, EXOs invade into β cells to arrest apoptosis and increase viability to preserve insulin secretion capacity [103]. Exosomes sourced from mesenchymal stem cells of the human umbilical cord were encapsulated in alginate/polyvinyl alcohol nano hydrogel to heal wounds that occurred due to diabetes. The applied nano hydrogel on diabetic rats promoted the proliferation, migration, and angiogenesis of mesenchymal stem cells and boosted the diabetic wound healing process. It also upregulated the level of VEGF *via* ERK1/2 pathway regulation. Altogether, the significant wound healing in rats could be due to the exosome-loaded nano hydrogel [104]. Another similar research has proven that exosomes sourced from pioglitazone-treated mesenchymal stem cells promoted migration, wound healing, and VEGF expression. They also enhanced p-AKT, p-eNOS, and p-PI3K protein expression and suppressed PTEN expression. The therapeutic potential of the exosomes augmented diabetic wound healing in diabetic rats by promoting angiogenesis [105]. Hence this research offers an exosome as a potential candidate for managing diabetic wounds.

Phytosome-based Nanocarriers

Phytosomes (PHYs) are molecular complexes of naturally active phytochemicals and phospholipids confined in their inner structures. Polyphenolic compounds are the essential part of phospholipids that exhibit better pharmacological action. The enhanced absorption and bioavailability are the potential advantages of PHYs. Furthermore, PHYs NPs augment the application of standard herbal extract [106]. The low bioavailability and low stability of natural antioxidant, rutin (RUT) were enhanced by encapsulating RUT in nano-PHYs. The therapeutic potency of nanophytosomes was evaluated in streptozotocin-injected diabetic rats. The orally administered RUT-loaded nano-PHYs (25 mg/kg/day for four weeks) regulated the liver marker enzyme activities and the level of glycated and total hemoglobin in the diabetic rats than the free RUT control group. Additionally, histopathological studies showed the restoration of diabetes-provoked complications in the pancreas, liver, and kidney by nano-PHYs. Hence, rutin-encapsulated PHYs were an effective technique in managing diabetic complications [107]. Kim *et al*. have prepared natural flavonoid-loaded phytosomes to examine the anti-diabetic effects on Type 2 diabetic mice. Chrysin (CHR) loaded PHYs were prepared by employing egg phospholipid at a molar ratio of 1:3. The administration of CHR-loaded PHYs significantly reduced the levels of fasting blood glucose, glucose tolerance, and insulin resistance in mice than the CHR-treated group than the control group. Besides, they also downregulated gluconeogenesis and stimulated the uptake of glucose in the treated mice's skeletal muscle. Hence, the nano-sized particles of CHR-loaded PHYs might be the reason for the enhanced bioavailability and greater antidiabetic effect [108].

Sphingosomes-based Nanocarriers

Sphingosomes (SPH) are concentric bilayer vesicles, mainly comprised of sphingolipids and cholesterol. Importantly, the sphingolipid plays a vital function in defining the biophysical properties of membrane, integrity, and topology in various signaling pathways including apoptosis and proliferation. Hence, the sphingolipid acts as a main player in diabetes. Their compatible nature and improved characteristics such as stable to acid hydrolysis, enhanced drug retention, therapeutic index, and flexibility to attain active targeting have made an efficient delivery system [109, 110]. Upcoming research has focused on the probable action of SPH in diabetes pathology and their involvement in the management of diabetes. Currently, no distinct SPH nanomedicine has been approved for diabetes management. The research in this intriguing area is still evolving for upcoming diabetic treatment.

Recent patents in the nanomedicine-based approach to managing diabetes are listed in Table **1** (https://patentscope.wipo.int/search/en/search.jsf).

Table 1. Patents on nanomedicine to manage diabetes.

Publication Number	Publication Date	Title	Applicants
WO/2022/168123	11.08.2022	Process for preparing nano-formulation for delivery of berbamine	Panjab university, chandigarh.
20220218788	14.07.2022	Orally delivered lipid nanoparticles target and reveal gut cd36 as a master regulator of systemic lipid homeostasis with differential gender responses	Northwestern university
WO/2022/136393	30.06.2022	Zein nanoparticles for use in hyperglycemic conditions	Universidad de navarra

(Table 1) cont.....

Publication Number	Publication Date	Title	Applicants
WO/2022/132102	23.06.2022	Method of manufacturing cafestol-loaded biopolymer nanoparticles	Ege üniversitesi
202241031530	17.06.2022	A process for preparing stable and biocompatible silver nanoparticles of Morinda citrifolia fruit extract for anti-diabetic activity	Adichunchanagiri University
201921026090	27.05.2022	Nanoparticles of Tinospora cordifolia (Giloy) stem with anti-hyperglycemic and anti-oxidant functions	Jiwaji University
201921026094	27.05.2022	Lecithin-berberine nanoparticles for diabetes	Jiwaji university
114209663	22.03.2022	Preparation and application of phloretin-loaded soybean lecithin-chitosan nanoparticles for preventing diabetes	Zhejiang University
3965746	16.03.2022	Orally delivered lipid nanoparticles target and reveal gut cd36 as a master regulator of systemic lipid homeostasis with differential gender responses	UNIV NORTHWESTERN
3952904	16.02.2022	Lipid-based nanoparticles and use of same in optimized insulin dosing regimens	SDG INC

CHALLENGES WITH NANOMEDICINE IN THE MANAGEMENT OF DIABETES

Persistent glycaemic control is a crucial determinant of long-term outcomes for patients with diabetes. The goal of management for type 1 and type 2 diabetes is maintaining blood glucose levels within healthy normoglycemic ranges. For type 1 diabetes patients, insulin replacement therapy is prescribed to mimic natural fluctuations in insulin levels throughout the day. For type 2 diabetes, initial treatment focuses on delaying disease progression through exercise and regulation of meals. Patients also receive oral and injectable medication that improves insulin production and function. However, insulin replacement therapy is often required as native insulin production diminishes. In case of diabetes mellitus, diabetic retinopathy can be utilized as a better option for therapy with the help of nanoparticles which can provide sustained delivery, targeted delivery to specific cells or tissues, improved delivery of both water-insoluble drugs and prominent biomolecule drugs, and reduced side effects, minimizing toxicological reactions. However, there can be unexpected issues, such as toxicity of the nanoparticles in the section of the eye, rupture of the inner part of the eye at the time of administration, and tolerability of the patients post administration which are also a factor for patients with diabetes. So, before the administration of the drug through retinopathy, the nanomedicines have to be characterized for better administration which will be acceptable to patients and provide better treatment [111].

CONCLUSION AND FUTURE PERSPECTIVES

Available antidiabetic therapeutic molecules exhibit their activity by enhancing glucose uptake or facilitating insulin release in peripheral tissues. Unfortunately, these molecules are ineffective in delaying or regulating hyperglycaemia leading to crucial insulin therapy. Conventionally, these molecules are often developed into tablets and other traditional oral dosage forms. Subsequently, these are associated with numerous side effects. These portray the incompetence of the available conventional therapeutic agents; notably, these dosage forms fail to deliver optimum concentrations at the desired site. Therefore, often higher doses are administered to obtain prolonged drug concentrations. Unequivocally, antidiabetic therapy aims to improve the patient's life precisely by delaying or averting the difficulties related to the disease, thereby lowering mortality. The prime focus of antidiabetic therapy is potentially to regulate and optimize glycaemic values. Maintaining glycaemic levels helps clinicians preclude further complications associated with diabetes, such as microvascular complications. A potential nano drug delivery system should effectively regulate the glucose levels precisely to a standard healthy value for a prolonged period. Various studies suggested that an adept delivery system, specifically when managing diabetic conditions, should be able to instantaneously adapt the drug release depending on blood glucose levels. Novel smart nanocarriers could potentially achieve this phenomenon. Besides, the nanosystems should attain stealth mode during prolonged systemic circulation, thus regulating glucose concentration. Enormous research have been conducted to develop novel glucose-sensitive patches loaded with varied nanosystems that could lead to better blood glucose regulation and further advancements in nanotechnology enabled researchers to develop biomedical devices that facilitate better disease management. Recent understanding of novel nanomedicines in managing other disease conditions should also percolate toward managing diabetes. Besides, these advancements should potentially eliminate the adverse effects associated with conventional therapies. Fascinatingly, there have been tremendous advancements in green chemistry/nanotechnology that may significantly aid in developing novel antidiabetic molecules. The current research outcomes involving nanotechnology and novel approaches to managing diabetes have encouraged researchers to explore this technology to its full potential.

REFERENCES

[1] Galicia-Garcia U, Benito-Vicente A, Jebari S, *et al.* Pathophysiology of Type 2 Diabetes Mellitus. Int J Mol Sci 2020; 21(17): 6275.
[http://dx.doi.org/10.3390/ijms21176275] [PMID: 32872570]

[2] Stumvoll M, Goldstein BJ, van Haeften TW. Type 2 diabetes: Principles of pathogenesis and therapy. Lancet 2005; 365(9467): 1333-46.
[http://dx.doi.org/10.1016/S0140-6736(05)61032-X] [PMID: 15823385]

[3] Chatterjee S, Khunti K, Davies MJ. Type 2 diabetes. Lancet 2017; 389(10085): 2239-51.
[http://dx.doi.org/10.1016/S0140-6736(17)30058-2] [PMID: 28190580]

[4] Chawla R, Chawla A, Jaggi S. Microvasular and macrovascular complications in diabetes mellitus: Distinct or continuum? Indian J Endocrinol Metab 2016; 20(4): 546-51.
[http://dx.doi.org/10.4103/2230-8210.183480] [PMID: 27366724]

[5] DeFronzo RA, Ferrannini E, Groop L, *et al.* Type 2 diabetes mellitus. Nat Rev Dis Primers 2015; 1(1): 15019.
[http://dx.doi.org/10.1038/nrdp.2015.19] [PMID: 27189025]

[6] Hamer JL. Hoosiers abroad--report of a case of Lassa fever. J Indiana State Med Assoc 1974; 67(7): 659-60.
[PMID: 4604539]

[7] Villena Gonzales W, Mobashsher A, Abbosh A. The progress of glucose monitoring—a review of invasive to minimally and non-invasive techniques, devices and sensors. Sensors 2019; 19(4): 800.
[http://dx.doi.org/10.3390/s19040800] [PMID: 30781431]

[8] Mago A, Tahir MJ, Khan MA, Ahmed KAHM, Munir MU. Nanomedicine: Advancement in healthcare. Ann Med Surg 2022; 79: 104078.
[http://dx.doi.org/10.1016/j.amsu.2022.104078] [PMID: 35812828]

[9] Prasad M, Lambe UP, Brar B, *et al.* Nanotherapeutics: An insight into healthcare and multi-dimensional applications in medical sector of the modern world. Biomed Pharmacother 2018; 97: 1521-37.
[http://dx.doi.org/10.1016/j.biopha.2017.11.026] [PMID: 29793315]

[10] Lemmerman LR, Das D, Higuita-Castro N, Mirmira RG, Gallego-Perez D. Nanomedicine-based strategies for diabetes: Diagnostics, monitoring, and treatment. Trends Endocrinol Metab 2020; 31(6): 448-58.
[http://dx.doi.org/10.1016/j.tem.2020.02.001] [PMID: 32396845]

[11] Reed J, Bain S, Kanamarlapudi V. A review of current trends with type 2 diabetes epidemiology, aetiology, pathogenesis, treatments and future perspectives. Diabetes Metab Syndr Obes 2021; 14: 3567-602.
[http://dx.doi.org/10.2147/DMSO.S319895] [PMID: 34413662]

[12] Kerry RG, Mahapatra GP, Maurya GK, *et al.* Molecular prospect of type-2 diabetes: Nanotechnology based diagnostics and therapeutic intervention. Rev Endocr Metab Disord 2021; 22(2): 421-51.
[http://dx.doi.org/10.1007/s11154-020-09606-0] [PMID: 33052523]

[13] He Y, Al-Mureish A, Wu N. Nanotechnology in the treatment of diabetic complications: A comprehensive narrative review. J Diabetes Res 2021; 2021: 1-11.
[http://dx.doi.org/10.1155/2021/6612063] [PMID: 34007847]

[14] Wang Y, Wang C, Li K, *et al.* Recent advances of nanomedicine-based strategies in diabetes and complications management: Diagnostics, monitoring, and therapeutics. J Control Release 2021; 330: 618-40.
[http://dx.doi.org/10.1016/j.jconrel.2021.01.002] [PMID: 33417985]

[15] Plaza-Oliver M, Santander-Ortega MJ, Lozano MV. Current approaches in lipid-based nanocarriers for oral drug delivery. Drug Deliv Transl Res 2021; 11(2): 471-97.

[http://dx.doi.org/10.1007/s13346-021-00908-7] [PMID: 33528830]

[16]   Guimarães D, Cavaco-Paulo A, Nogueira E. Design of liposomes as drug delivery system for therapeutic applications. Int J Pharm 2021; 601: 120571.
[http://dx.doi.org/10.1016/j.ijpharm.2021.120571] [PMID: 33812967]

[17]   Sarhadi S, Moosavian SA, Mashreghi M, *et al.* B12-functionalized PEGylated liposomes for the oral delivery of insulin: *In vitro* and *in vivo* studies. J Drug Deliv Sci Technol 2022; 69: 103141.
[http://dx.doi.org/10.1016/j.jddst.2022.103141]

[18]   Wang Q, Wei C, Weng W, *et al.* Enhancement of oral bioavailability and hypoglycemic activity of liquiritin-loaded precursor liposome. Int J Pharm 2021; 592: 120036.
[http://dx.doi.org/10.1016/j.ijpharm.2020.120036] [PMID: 33152478]

[19]   Muneer S, Masood Z, Anjum S. Proliposomes as pharmaceutical drug delivery system: A brief review. J Text Sci Eng 2017; 08(03).

[20]   Khan S, Madni A, Rahim MA, *et al.* Enhanced *in vitro* release and permeability of glibenclamide by proliposomes: Development, characterization and histopathological evaluation. J Drug Deliv Sci Technol 2021; 63: 102450.
[http://dx.doi.org/10.1016/j.jddst.2021.102450]

[21]   Elkarray SM, Farid RM, Abd-Alhaseeb MM, Omran GA, Habib DA. Intranasal repaglinide-solid lipid nanoparticles integrated *in situ* gel outperform conventional oral route in hypoglycemic activity. J Drug Deliv Sci Technol 2022; 68: 103086.
[http://dx.doi.org/10.1016/j.jddst.2021.103086]

[22]   El-Salamouni NS, Gowayed MA, Seiffein NL, Abdel- Moneim RA, Kamel MA, Labib GS. Valsartan solid lipid nanoparticles integrated hydrogel: A challenging repurposed use in the treatment of diabetic foot ulcer, *in-vitro/ in-vivo* experimental study. Int J Pharm 2021; 592: 120091.
[http://dx.doi.org/10.1016/j.ijpharm.2020.120091] [PMID: 33197564]

[23]   Elmowafy M, Al-Sanea MM. Nanostructured lipid carriers (NLCs) as drug delivery platform: Advances in formulation and delivery strategies. Saudi Pharm J 2021; 29(9): 999-1012.
[http://dx.doi.org/10.1016/j.jsps.2021.07.015] [PMID: 34588846]

[24]   Kenechukwu FC, Isaac GT, Nnamani DO, Momoh MA, Attama AA. Enhanced circulation longevity and pharmacodynamics of metformin from surface-modified nanostructured lipid carriers based on solidified reverse micellar solutions. Heliyon 2022; 8(3): e09100.
[http://dx.doi.org/10.1016/j.heliyon.2022.e09100] [PMID: 35313488]

[25]   Piazzini V, Micheli L, Luceri C, *et al.* Nanostructured lipid carriers for oral delivery of silymarin: Improving its absorption and *in vivo* efficacy in type 2 diabetes and metabolic syndrome model. Int J Pharm 2019; 572: 118838.
[http://dx.doi.org/10.1016/j.ijpharm.2019.118838] [PMID: 31715362]

[26]   Ozogul Y, Karsli GT, Durmuş M, *et al.* Recent developments in industrial applications of nanoemulsions. Adv Colloid Interface Sci 2022; 304: 102685.
[http://dx.doi.org/10.1016/j.cis.2022.102685] [PMID: 35504214]

[27]   Razavi R, Kenari RE, Farmani J, Jahanshahi M. Preparation of double-layer nanoemulsions with controlled release of glucose as prevention of hypoglycemia in diabetic patients. Biomed Pharmacother 2021; 138: 111464.
[http://dx.doi.org/10.1016/j.biopha.2021.111464] [PMID: 33725590]

[28]   Hussein J, El-Naggar ME. Synthesis of an environmentally quercetin nanoemulsion to ameliorate diabetic-induced cardiotoxicity. Biocatal Agric Biotechnol 2021; 33: 101983.
[http://dx.doi.org/10.1016/j.bcab.2021.101983]

[29]   Zhu T, Kang W, Yang H, *et al.* Advances of microemulsion and its applications for improved oil recovery. Adv Colloid Interface Sci 2022; 299: 102527.
[http://dx.doi.org/10.1016/j.cis.2021.102527] [PMID: 34607652]

[30] Momoh MA, Franklin KC, Agbo CP, *et al.* Microemulsion-based approach for oral delivery of insulin: formulation design and characterization. Heliyon 2020; 6(3): e03650.
[http://dx.doi.org/10.1016/j.heliyon.2020.e03650] [PMID: 32258491]

[31] Kaur I, Nallamothu B, Kuche K, Katiyar SS, Chaudhari D, Jain S. Exploring protein stabilized multiple emulsion with permeation enhancer for oral delivery of insulin. Int J Biol Macromol 2021; 167: 491-501.
[http://dx.doi.org/10.1016/j.ijbiomac.2020.11.190] [PMID: 33279562]

[32] Dabholkar N, Waghule T, Krishna Rapalli V, *et al.* Lipid shell lipid nanocapsules as smart generation lipid nanocarriers. J Mol Liq 2021; 339: 117145.
[http://dx.doi.org/10.1016/j.molliq.2021.117145]

[33] El-Hussien D, El-Zaafarany GM, Nasr M, Sammour O. Chrysin nanocapsules with dual anti-glycemic and anti-hyperlipidemic effects: Chemometric optimization, physicochemical characterization and pharmacodynamic assessment. Int J Pharm 2021; 592: 120044.
[http://dx.doi.org/10.1016/j.ijpharm.2020.120044] [PMID: 33157212]

[34] Shamekhi F, Tamjid E, Khajeh K. Development of chitosan coated calcium-alginate nanocapsules for oral delivery of liraglutide to diabetic patients. Int J Biol Macromol 2018; 120(Pt A): 460-7.
[http://dx.doi.org/10.1016/j.ijbiomac.2018.08.078] [PMID: 30125628]

[35] Akhtar N, Mohammed SAA, Khan RA, *et al.* Self-Generating nano-emulsification techniques for alternatively-routed, bioavailability enhanced delivery, especially for anti-cancers, anti-diabetics, and miscellaneous drugs of natural, and synthetic origins. J Drug Deliv Sci Technol 2020; 58: 101808.
[http://dx.doi.org/10.1016/j.jddst.2020.101808]

[36] Bravo-Alfaro DA, Muñoz-Correa MOF, Santos-Luna D, *et al.* Encapsulation of an insulin-modified phosphatidylcholine complex in a self-nanoemulsifying drug delivery system (SNEDDS) for oral insulin delivery. J Drug Deliv Sci Technol 2020; 57: 101622.
[http://dx.doi.org/10.1016/j.jddst.2020.101622]

[37] Singh D, Bedi N, Tiwary AK, Kurmi BD, Bhattacharya S. Natural bio functional lipids containing solid self-microemulsifying drug delivery system of Canagliflozin for synergistic prevention of type 2 diabetes mellitus. J Drug Deliv Sci Technol 2022; 69: 103138.
[http://dx.doi.org/10.1016/j.jddst.2022.103138]

[38] Wong KH, Lu A, Chen X, Yang Z. Natural Ingredient-Based Polymeric Nanoparticles for Cancer Treatment. Molecules 2020; 25(16): 3620.
[http://dx.doi.org/10.3390/molecules25163620] [PMID: 32784890]

[39] Mohammed M, Syeda J, Wasan K, Wasan E. An Overview of Chitosan Nanoparticles and Its Application in Non-Parenteral Drug Delivery. Pharmaceutics 2017; 9(4): 53.
[http://dx.doi.org/10.3390/pharmaceutics9040053] [PMID: 29156634]

[40] Lari AS, Zahedi P, Ghourchian H, Khatibi A. Microfluidic-based synthesized carboxymethyl chitosan nanoparticles containing metformin for diabetes therapy: *In vitro* and *in vivo* assessments. Carbohydr Polym 2021; 261: 117889.
[http://dx.doi.org/10.1016/j.carbpol.2021.117889] [PMID: 33766375]

[41] Abdel-Moneim A, El-Shahawy A, Yousef AI, Abd El-Twab SM, Elden ZE, Taha M. Novel polydatin-loaded chitosan nanoparticles for safe and efficient type 2 diabetes therapy: In silico, *in vitro* and *in vivo* approaches. Int J Biol Macromol 2020; 154: 1496-504.
[http://dx.doi.org/10.1016/j.ijbiomac.2019.11.031] [PMID: 31758992]

[42] Choukaife H, Doolaanea AA, Alfatama M. Alginate Nanoformulation: Influence of Process and Selected Variables. Pharmaceuticals 2020; 13(11): 335.
[http://dx.doi.org/10.3390/ph13110335] [PMID: 33114120]

[43] Kumar D, Gautam A, Rohatgi S, Kundu PP. Synthesis of vildagliptin loaded acrylamide- g-psyllium/alginate-based core-shell nanoparticles for diabetes treatment. Int J Biol Macromol 2022;

218: 82-93.
[http://dx.doi.org/10.1016/j.ijbiomac.2022.07.066] [PMID: 35841963]

[44] Shaedi N, Naharudin I, Choo CY, Wong TW. Design of oral intestinal-specific alginate-vitexin nanoparticulate system to modulate blood glucose level of diabetic rats. Carbohydr Polym 2021; 254: 117312.
[http://dx.doi.org/10.1016/j.carbpol.2020.117312] [PMID: 33357875]

[45] Hu Q, Lu Y, Luo Y. Recent advances in dextran-based drug delivery systems: From fabrication strategies to applications. Carbohydr Polym 2021; 264: 117999.
[http://dx.doi.org/10.1016/j.carbpol.2021.117999] [PMID: 33910733]

[46] Bao X, Qian K, Yao P. Insulin- and cholic acid-loaded zein/casein–dextran nanoparticles enhance the oral absorption and hypoglycemic effect of insulin. J Mater Chem B Mater Biol Med 2021; 9(31): 6234-45.
[http://dx.doi.org/10.1039/D1TB00806D] [PMID: 34328161]

[47] Jamwal S, Ram B, Ranote S, Dharela R, Chauhan GS. New glucose oxidase-immobilized stimuli-responsive dextran nanoparticles for insulin delivery. Int J Biol Macromol 2019; 123: 968-78.
[http://dx.doi.org/10.1016/j.ijbiomac.2018.11.147] [PMID: 30448487]

[48] Hamid Akash MS, Rehman K, Chen S. Natural and synthetic polymers as drug carriers for delivery of therapeutic proteins. Polym Rev 2015; 55(3): 371-406.
[http://dx.doi.org/10.1080/15583724.2014.995806]

[49] Ren T, Zheng X, Bai R, Yang Y, Jian L. Utilization of PLGA nanoparticles in yeast cell wall particle system for oral targeted delivery of exenatide to improve its hypoglycemic efficacy. Int J Pharm 2021; 601: 120583.
[http://dx.doi.org/10.1016/j.ijpharm.2021.120583] [PMID: 33839225]

[50] Maity S, Chakraborti AS. Formulation, physico-chemical characterization and antidiabetic potential of naringenin-loaded poly D, L lactide-co-glycolide (N-PLGA) nanoparticles. Eur Polym J 2020; 134: 109818.
[http://dx.doi.org/10.1016/j.eurpolymj.2020.109818]

[51] Casalini T, Rossi F, Castrovinci A, Perale G. A perspective on polylactic acid-based polymers use for nanoparticles synthesis and applications. Front Bioeng Biotechnol 2019; 7: 259.
[http://dx.doi.org/10.3389/fbioe.2019.00259] [PMID: 31681741]

[52] El-Naggar ME, Al-Joufi F, Anwar M, Attia MF, El-Bana MA. Curcumin-loaded PLA-PEG copolymer nanoparticles for treatment of liver inflammation in streptozotocin-induced diabetic rats. Colloids Surf B Biointerfaces 2019; 177: 389-98.
[http://dx.doi.org/10.1016/j.colsurfb.2019.02.024] [PMID: 30785036]

[53] Jacob S, Nair AB, Shah J. Emerging role of nanosuspensions in drug delivery systems. Biomater Res 2020; 24(1): 3.
[http://dx.doi.org/10.1186/s40824-020-0184-8] [PMID: 31969986]

[54] Gaur PK. Nanosuspension of flavonoid-rich fraction from *psidium guajava* Linn for improved type 2-diabetes potential. J Drug Deliv Sci Technol 2021; 62: 102358.
[http://dx.doi.org/10.1016/j.jddst.2021.102358]

[55] Hemalatha S , Monisha J . Formulation and evaluation of anti-diabetic activity of repaglinide nanosuspension. Int J Pharml Res Life Sci 2020; 8(2): 17-21.
[http://dx.doi.org/10.26452/ijprls.v8i2.1208]

[56] Sampathkumar SG, Yarema KJ. Dendrimers in cancer treatment and diagnosis. Nanotechnol Sci. Weinheim, Germany: Wiley-VCH Verlag GmbH & Co. KGaA; 2007.

[57] Sherje AP, Jadhav M, Dravyakar BR, Kadam D. Dendrimers: A versatile nanocarrier for drug delivery and targeting. Int J Pharm 2018; 548(1): 707-20.
[http://dx.doi.org/10.1016/j.ijpharm.2018.07.030] [PMID: 30012508]

[58] Kim Y, Park EJ, Na DH. Recent progress in dendrimer-based nanomedicine development. Arch Pharm Res 2018; 41(6): 571-82.
[http://dx.doi.org/10.1007/s12272-018-1008-4] [PMID: 29450862]

[59] Akhtar S, Chandrasekhar B, Yousif MHM, Renno W, Benter IF, El-Hashim AZ. Chronic administration of nano-sized PAMAM dendrimers *in vivo* inhibits EGFR-ERK1/2-ROCK signaling pathway and attenuates diabetes-induced vascular remodeling and dysfunction. Nanomedicine 2019; 18: 78-89.
[http://dx.doi.org/10.1016/j.nano.2019.02.012] [PMID: 30844576]

[60] Zhang D, Huang Q. Encapsulation of astragaloside with matrix metalloproteinase-2-responsive hyaluronic acid end-conjugated polyamidoamine dendrimers improves wound healing in diabetes. J Biomed Nanotechnol 2020; 16(8): 1229-40.
[http://dx.doi.org/10.1166/jbn.2020.2971] [PMID: 33397553]

[61] Milovanovic M, Arsenijevic A, Milovanovic J, Kanjevac T, Arsenijevic N. Nanoparticles in antiviral therapy. Antimicrobial Nanoarchitectonics. Elsevier 2017; pp. 383-410.
[http://dx.doi.org/10.1016/B978-0-323-52733-0.00014-8]

[62] Xin X, Chen J, Chen L, Wang J, Liu X, Chen F. Lyophilized insulin micelles for long-term storage and regulation of blood glucose for preventing hypoglycemia. Chem Eng J 2022; 435: 134929.
[http://dx.doi.org/10.1016/j.cej.2022.134929]

[63] Wen N, Lü S, Xu X, *et al.* A polysaccharide-based micelle-hydrogel synergistic therapy system for diabetes and vascular diabetes complications treatment. Mater Sci Eng C 2019; 100: 94-103.
[http://dx.doi.org/10.1016/j.msec.2019.02.081] [PMID: 30948130]

[64] Fuchs S, Shariati K, Ma M. Specialty tough hydrogels and their biomedical applications. Adv Healthc Mater 2020; 9(2): 1901396.
[http://dx.doi.org/10.1002/adhm.201901396] [PMID: 31846228]

[65] Zou M, Chi J, Jiang Z, *et al.* Functional thermosensitive hydrogels based on chitin as RIN-m5F cell carrier for the treatment of diabetes. Int J Biol Macromol 2022; 206: 453-66.
[http://dx.doi.org/10.1016/j.ijbiomac.2022.02.175] [PMID: 35247418]

[66] Patel J, Maiti S, Moorthy NSHN. Repaglinide-laden hydrogel particles of xanthan gum derivatives for the management of diabetes. Carbohydr Polym 2022; 287: 119354.
[http://dx.doi.org/10.1016/j.carbpol.2022.119354] [PMID: 35422303]

[67] Nagaraju K, Reddy R, Reddy N. A review on protein functionalized carbon nanotubes. J Appl Biomater Funct Mater 2015; 13(4): 301-12.
[http://dx.doi.org/10.5301/jabfm.5000231] [PMID: 26660626]

[68] Ma J, Liu J, Lu CW, Cai DF. Pachymic acid modified carbon nanoparticles reduced angiogenesis *via* inhibition of MMP-3. Int J Clin Exp Pathol 2015; 8(5): 5464-70.
[PMID: 26191251]

[69] Ding X, Su Y, Wang C, *et al.* Synergistic suppression of tumor angiogenesis by the co-delivering of vascular endothelial growth factor targeted sirna and candesartan mediated by functionalized carbon nanovectors. ACS Appl Mater Interfaces 2017; 9(28): 23353-69.
[http://dx.doi.org/10.1021/acsami.7b04971] [PMID: 28617574]

[70] Villena Gonzales W, Mobashsher A, Abbosh A. The progress of glucose monitoring—a review of invasive to minimally and non-invasive techniques, devices and sensors. Sensors 2019; 19(4): 800.
[http://dx.doi.org/10.3390/s19040800] [PMID: 30781431]

[71] Bruen D, Delaney C, Florea L, Diamond D. Glucose Sensing for Diabetes Monitoring: Recent Developments. Sensors 2017; 17(8): 1866.
[http://dx.doi.org/10.3390/s17081866] [PMID: 28805693]

[72] Barone PW, Baik S, Heller DA, Strano MS. Near-infrared optical sensors based on single-walled carbon nanotubes. Nat Mater 2005; 4(1): 86-92.

[http://dx.doi.org/10.1038/nmat1276] [PMID: 15592477]

[73]   Yum K, Ahn JH, McNicholas TP, *et al.* Boronic acid library for selective, reversible near-infrared fluorescence quenching of surfactant suspended single-walled carbon nanotubes in response to glucose. ACS Nano 2012; 6(1): 819-30.
[http://dx.doi.org/10.1021/nn204323f] [PMID: 22133474]

[74]   Abdal Dayem A, Hossain M, Lee S, *et al.* The Role of Reactive Oxygen Species (ROS) in the Biological Activities of Metallic Nanoparticles. Int J Mol Sci 2017; 18(1): 120.
[http://dx.doi.org/10.3390/ijms18010120] [PMID: 28075405]

[75]   Majdalawieh A, Kanan MC, El-Kadri O, Kanan SM. Recent advances in gold and silver nanoparticles: synthesis and applications. J Nanosci Nanotechnol 2014; 14(7): 4757-80.
[http://dx.doi.org/10.1166/jnn.2014.9526] [PMID: 24757945]

[76]   Katsumi H, Fukui K, Sato K, *et al.* Pharmacokinetics and preventive effects of platinum nanoparticles as reactive oxygen species scavengers on hepatic ischemia/reperfusion injury in mice. Metallomics 2014; 6(5): 1050-6.
[http://dx.doi.org/10.1039/C4MT00018H] [PMID: 24658875]

[77]   Fan J, Yin JJ, Ning B, *et al.* Direct evidence for catalase and peroxidase activities of ferritin–platinum nanoparticles. Biomaterials 2011; 32(6): 1611-8.
[http://dx.doi.org/10.1016/j.biomaterials.2010.11.004] [PMID: 21112084]

[78]   Bhardwaj M, Yadav P, Dalal S, Kataria SK. A review on ameliorative green nanotechnological approaches in diabetes management. Biomed Pharmacother 2020; 127: 110198.
[http://dx.doi.org/10.1016/j.biopha.2020.110198] [PMID: 32559845]

[79]   Saratale GD, Saratale RG, Benelli G, *et al.* Anti-diabetic potential of silver nanoparticles synthesized with argyreia nervosa leaf extract high synergistic antibacterial activity with standard antibiotics against foodborne bacteria. J Cluster Sci 2017; 28(3): 1709-27.
[http://dx.doi.org/10.1007/s10876-017-1179-z]

[80]   Saratale RG, Shin HS, Kumar G, Benelli G, Kim DS, Saratale GD. Exploiting antidiabetic activity of silver nanoparticles synthesized using *Punica granatum* leaves and anticancer potential against human liver cancer cells (HepG2). Artif Cells Nanomed Biotechnol 2018; 46(1): 211-22.
[http://dx.doi.org/10.1080/21691401.2017.1337031] [PMID: 28612655]

[81]   Vijaya Sankar M AS. *In-vitro* screening of antidiabetic and antimicrobial activity against green synthesized AgNO$_3$ using seaweeds. J Nanomed Nanotechnol 2015; s6: s6.
[http://dx.doi.org/10.4172/2157-7439.S6-001]

[82]   Vardatsikos G, Pandey NR, Srivastava AK. Insulino-mimetic and anti-diabetic effects of zinc. J Inorg Biochem 2013; 120: 8-17.
[http://dx.doi.org/10.1016/j.jinorgbio.2012.11.006] [PMID: 23266931]

[83]   Kim S, Jung Y, Kim D, Koh H, Chung J. Extracellular zinc activates p70 S6 kinase through the phosphatidylinositol 3-kinase signaling pathway. J Biol Chem 2000; 275(34): 25979-84.
[http://dx.doi.org/10.1074/jbc.M001975200] [PMID: 10851233]

[84]   Eom SJ, Kim EY, Lee JE, *et al.* Zn($2+$) induces stimulation of the c-Jun N-terminal kinase signaling pathway through phosphoinositide 3-Kinase. Mol Pharmacol 2001; 59(5): 981-6.
[http://dx.doi.org/10.1124/mol.59.5.981] [PMID: 11306679]

[85]   Venkatachalam M, Govindaraju K, Mohamed Sadiq A, Tamilselvan S, Ganesh Kumar V, Singaravelu G. Functionalization of gold nanoparticles as antidiabetic nanomaterial. Spectrochim Acta A Mol Biomol Spectrosc 2013; 116: 331-8.
[http://dx.doi.org/10.1016/j.saa.2013.07.038] [PMID: 23973575]

[86]   Dhas TS, Kumar VG, Karthick V, Vasanth K, Singaravelu G, Govindaraju K. Effect of biosynthesized gold nanoparticles by *sargassum swartzii* in alloxan induced diabetic rats. Enzyme Microb Technol 2016; 95: 100-6.

[http://dx.doi.org/10.1016/j.enzmictec.2016.09.003] [PMID: 27866603]

[87]   Karthick V, Kumar VG, Dhas TS, Singaravelu G, Sadiq AM, Govindaraju K. Effect of biologically synthesized gold nanoparticles on alloxan-induced diabetic rats—An *in vivo* approach. Colloids Surf B Biointerfaces 2014; 122: 505-11.
[http://dx.doi.org/10.1016/j.colsurfb.2014.07.022] [PMID: 25092583]

[88]   Guo Y, Jiang N, Zhang L, Yin M. Green synthesis of gold nanoparticles from *fritillaria cirrhosa* and its anti-diabetic activity on streptozotocin induced rats. Arab J Chem 2020; 13(4): 5096-106.
[http://dx.doi.org/10.1016/j.arabjc.2020.02.009]

[89]   Miñon-Hernández D, Villalobos-Espinosa J, Santiago-Roque I, *et al.* Biofunctionality of native and nano-structured blue corn starch in prediabetic Wistar rats. CYTA J Food 2018; 16(1): 477-83.
[http://dx.doi.org/10.1080/19476337.2017.1422279]

[90]   Sharma G, Sharma AR, Nam JS, Doss GPC, Lee SS, Chakraborty C. Nanoparticle based insulin delivery system: the next generation efficient therapy for Type 1 diabetes. J Nanobiotechnology 2015; 13(1): 74.
[http://dx.doi.org/10.1186/s12951-015-0136-y] [PMID: 26498972]

[91]   Li X, Szewczuk M, Malardier-Jugroot C. Folic acid-conjugated amphiphilic alternating copolymer as a new active tumor targeting drug delivery platform. Drug Des Devel Ther 2016; 10: 4101-10.
[http://dx.doi.org/10.2147/DDDT.S123386] [PMID: 28008233]

[92]   Yoo J, Park C, Yi G, Lee D, Koo H. Active targeting strategies using biological ligands for nanoparticle drug delivery systems. Cancers 2019; 11(5): 640.
[http://dx.doi.org/10.3390/cancers11050640] [PMID: 31072061]

[93]   Kaasalainen M, Rytkönen J, Mäkilä E, Närvänen A, Salonen J. Electrostatic interaction on loading of therapeutic peptide GLP-1 into porous silicon nanoparticles. Langmuir 2015; 31(5): 1722-9.
[http://dx.doi.org/10.1021/la5047047] [PMID: 25604519]

[94]   Araújo F, Shrestha N, Shahbazi MA, *et al.* The impact of nanoparticles on the mucosal translocation and transport of GLP-1 across the intestinal epithelium. Biomaterials 2014; 35(33): 9199-207.
[http://dx.doi.org/10.1016/j.biomaterials.2014.07.026] [PMID: 25109441]

[95]   Martinelli C, Pucci C, Ciofani G. Nanostructured carriers as innovative tools for cancer diagnosis and therapy. APL Bioeng 2019; 3(1): 011502.
[http://dx.doi.org/10.1063/1.5079943] [PMID: 31069332]

[96]   Pérez-Ortiz M, Zapata-Urzúa C, Acosta GA, Álvarez-Lueje A, Albericio F, Kogan MJ. Gold nanoparticles as an efficient drug delivery system for GLP-1 peptides. Colloids Surf B Biointerfaces 2017; 158: 25-32.
[http://dx.doi.org/10.1016/j.colsurfb.2017.06.015] [PMID: 28662391]

[97]   Jain A, Jain SK. l-Valine appended PLGA nanoparticles for oral insulin delivery. Acta Diabetol 2015; 52(4): 663-76.
[http://dx.doi.org/10.1007/s00592-015-0714-3] [PMID: 25655131]

[98]   de Vrueh RL, Smith PL, Lee CP. Transport of L-valine-acyclovir *via* the oligopeptide transporter in the human intestinal cell line, Caco-2. J Pharmacol Exp Ther 1998; 286(3): 1166-70.
[PMID: 9732374]

[99]   Martins JP, Figueiredo P, Wang S, *et al.* Neonatal fc receptor-targeted lignin-encapsulated porous silicon nanoparticles for enhanced cellular interactions and insulin permeation across the intestinal epithelium. Bioact Mater 2022; 9: 299-315.
[http://dx.doi.org/10.1016/j.bioactmat.2021.08.007] [PMID: 34820572]

[100]   Lutz H, Hu S, Dinh PU, Cheng K. Cells and cell derivatives as drug carriers for targeted delivery. Med Drug Discov 2019; 3: 100014.
[http://dx.doi.org/10.1016/j.medidd.2020.100014]

[101]   Hamidi M, Tajerzadeh H. Carrier erythrocytes: An overview. Drug Deliv 2003; 10(1): 9-20.

[http://dx.doi.org/10.1080/713840329] [PMID: 12554359]

[102] Xu X, Xu Y, Li Y, *et al.* Glucose-responsive erythrocyte-bound nanoparticles for continuously modulated insulin release. Nano Res 2022; 15(6): 5205-15.
[http://dx.doi.org/10.1007/s12274-022-4105-0]

[103] Sun Y, Tao Q, Wu X, Zhang L, Liu Q, Wang L. The utility of exosomes in diagnosis and therapy of diabetes mellitus and associated complications. Front Endocrinol 2021; 12: 756581.
[http://dx.doi.org/10.3389/fendo.2021.756581] [PMID: 34764939]

[104] Zhang Y, Zhang P, Gao X, Chang L, Chen Z, Mei X. Preparation of exosomes encapsulated nanohydrogel for accelerating wound healing of diabetic rats by promoting angiogenesis. Mater Sci Eng C 2021; 120: 111671.
[http://dx.doi.org/10.1016/j.msec.2020.111671] [PMID: 33545836]

[105] Hu Y, Tao R, Chen L, *et al.* Exosomes derived from pioglitazone-pretreated MSCs accelerate diabetic wound healing through enhancing angiogenesis. J Nanobiotechnology 2021; 19(1): 150.
[http://dx.doi.org/10.1186/s12951-021-00894-5] [PMID: 34020670]

[106] Barani M, Sangiovanni E, Angarano M, *et al.* Phytosomes as innovative delivery systems for phytochemicals: A comprehensive review of literature. Int J Nanomedicine 2021; 16: 6983-7022.
[http://dx.doi.org/10.2147/IJN.S318416] [PMID: 34703224]

[107] Amjadi S, Shahnaz F, Shokouhi B, *et al.* Nanophytosomes for enhancement of rutin efficacy in oral administration for diabetes treatment in streptozotocin-induced diabetic rats. Int J Pharm 2021; 610: 121208.
[http://dx.doi.org/10.1016/j.ijpharm.2021.121208] [PMID: 34673162]

[108] Kim S, Imm JY. The effect of chrysin-loaded phytosomes on insulin resistance and blood sugar control in type 2 diabetic db/db Mice. Molecules 2020; 25(23): 5503.
[http://dx.doi.org/10.3390/molecules25235503] [PMID: 33255372]

[109] Hannun YA, Obeid LM. Principles of bioactive lipid signalling: Lessons from sphingolipids. Nat Rev Mol Cell Biol 2008; 9(2): 139-50.
[http://dx.doi.org/10.1038/nrm2329] [PMID: 18216770]

[110] Russo SB, Ross JS, Cowart LA. Sphingolipids in obesity, type 2 diabetes, and metabolic disease. Handb Exp Pharmacol. 2013; p. 373.401

[111] Pickup JC. Management of diabetes mellitus: Is the pump mightier than the pen? Nat Rev Endocrinol 2012; 8(7): 425-33.
[http://dx.doi.org/10.1038/nrendo.2012.28] [PMID: 22371161]

CHAPTER 5

Nanoscience for Nucleotide Delivery in Diabetes

Ali Rastegari[1,*]

[1] *Department of Pharmaceutics and Pharmaceutical Nanotechnology, School of Pharmacy, Iran University of Medical Sciences, Tehran, Iran*

Abstract: The convergence of nanoscience and nucleotide delivery holds tremendous promise in revolutionizing diabetes treatment. Nucleotide delivery emerged as a promising tool to modulate gene expression and cellular function in diabetes. Integration of nanoscience and nucleotide delivery in diabetes treatment opens avenues for efficient therapies. This approach has the potential to significantly improve glucose regulation and mitigate long-term complications associated with the disease. This chapter discussed DNA and RNA delivery approaches in diabetes treatment and the future and challenges of nucleotide delivery in diabetes.

Keywords: Delivery, Diabetes, Gene, Nanotechnology.

INTRODUCTION

Current treatments for diabetes often rely on insulin injections, oral medications, and lifestyle changes. However, gene therapy has emerged as a cutting-edge approach that has the potential to provide long-lasting solutions to this global health epidemic. Studies have shown that diabetes disease could be related to several genes [1, 2]. Furthermore, protein and small molecules delivery are limited and cannot be used for the treatment of every condition of disease. However, accordingly RNA and DNA are precursors of proteins, they can be used as a promising approach to the treatment of different diseases. Nucleotide delivery even can be used for gene editing of host's DNA to cure a genetic defect as opposed to just providing a simple treatment [3, 4]. Nucleotide delivery is defined as the delivery of genetic material including DNA plasmid, or RNA into the cell for production of desired proteins or inhibiting protein expression to correct or modulate a disease. Nucleic acids have a highly negative charge and their intracellular uptake is limited due to the presence of the force of repulsion between nucleic acids and the negatively charged plasma membrane. Furthermore, nucleic acids are rapidly cleared from the body due to degradation

---

* **Corresponding author Ali Rastegari:** Department of Pharmaceutics and Pharmaceutical Nanotechnology, School of Pharmacy, Iran University of Medical Sciences, Tehran, Iran; E-mail: rategari.a@iums.ac.ir

by endonucleases [5]. In this regard, studies have shown that cationic nanoparticles act as a powerful carrier for the protection of nucleic acids from degradation and also enhance the transfection efficiency and gene expression into the targeted tissue (Fig. **1**) [6].

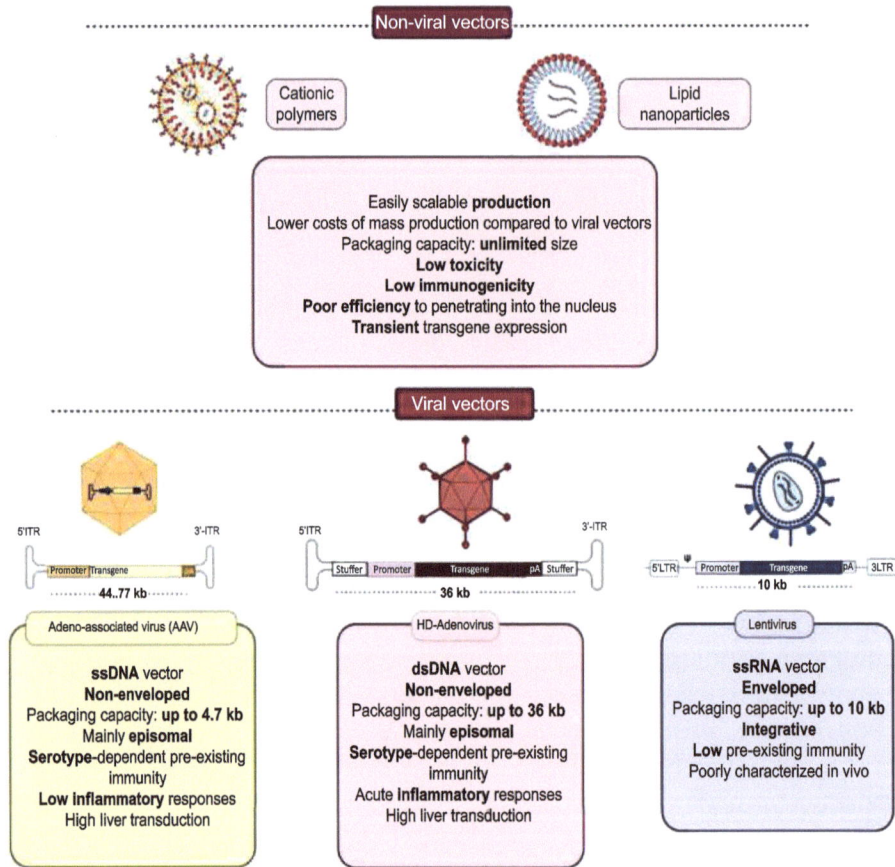

Fig. (1). Non-viral and viral vectors for nucleotide delivery [7].

Nanoscience has revolutionized the field of gene therapy, offering promising solutions to tackle complex diseases like diabetes. The integration of nanotechnology and gene therapy holds immense potential in transforming diabetes treatment [8].

The use of nanoparticles for nucleotide delivery could efficiently protect the degradation of nucleic acids and based on their chemical structure, increase nucleic acid cellular uptake and endosomal escape. In general, cationic nanoparticles based on polymers or lipids will be used to electrostatically condense with the nucleic acid with negative charged [9]. The positive charge of nanoparticles is usually achieved by using amine groups in their structures which will be protonated at physiological pH (pKa ~7.4). Many studies investigated synthetic and natural polymers for nucleotide delivery, for example, chitosan, poly-L-lysine and polyethyleneimine [10, 11]. As mentioned in previous chapters, the use of conventional therapeutic agents and small molecule delivery for diabetes treatment and control of blood glucose has several limitations. Accordingly, newer physiological approaches like nucleotide delivery could be a good candidate for the treatment of diabetes. In this chapter, we briefly discussed DNA and RNA delivery by using nanoparticles for the treatment of different types of diabetes.

DNA DELIVERY APPROACH

Plasmid DNA can encode information for the expression of therapeutic proteins in different diseases. In one study, biodegradable poly [α-(4-aminobutyl)-L-glycolic acid] (PAGA) could efficiently protect plasmid DNA pCAGGS from degradation and reduce the development of insulitis in non-obese diabetic (NOD) mice. Their results have shown that using this polymeric nanoparticle could increase the stability of plasmid DNA from 10 minutes to 60 minutes and make serum mIL-10 level peak at 5 days which could be detectable for 9 weeks. The study showed that using PAGA/plasmid DNA complex could prevent autoimmune diabetes. This formulation significantly decreased severe insulitis in NOD mice, 15.7% insulitis in treated group compared with 90.9% in non-treated group [12]. In other study, researchers used cationic nanoparticles by blending lactide-co-glycolide (PLGA) and methacrylate copolymer (Eudragit® E100) to deliver a therapeutic DNA encoding mouse interleukin-10 in the muscle of mice. Their results have shown that the prepared nanoparticles could effectively escape from the endosome and the transfection efficiency was significantly higher than PLGA nanoparticles. Elevation of interleukin-10 level can facilitate the suppression of interferon-gamma levels, which can reduce islet infiltration. By muscular injection of cationic nanoparticles containing DNA plasmid IL-10, a lower blood glucose level was achieved compared with alone plasmid and histological assessment showed no chronic inflammatory responses in the muscles [13]. Studies demonstrated that muscular injection could be an efficient route for gene delivery due to good accessibility, and vascularization, which make it as a suitable route for gene delivery to make long-lasting protein expression [14 - 16].

As mentioned previously, glucagon-like peptide-1 (GLP-1) is a treatment option in diabetes. Researchers are trying to produce GLP-1 endogenously by using GLP-1 plasmid to diminish the injection of GLP-1 in diabetic patients as a conventional therapy. GLP-1 has a very short half-life due to rapid degradation and after subcutaneous injection, its concentrations remain in therapeutic window just for 2 hours [6, 17]. In one study, the GLP-1 gene delivery was performed using a chicken β-actin promoter (pβGLP1). Their results have shown the GLP-1 mRNA level was increased after 24 hours of transfection which showed a dose-dependent response.

In their study, polyethyleneimine (PEI) was used as a cationic polymer for carrying and protecting DNA plasmid. Their results have shown that after a single injection of polyethyleneimine/pβGLP1 complex into diabetic rats, the insulin secretion was elevated and therefore the blood glucose levels decreased, and this effect was maintained for 2 weeks [18]. According to the requirement of active form of GLP-1 secretion in response to blood glucose, researchers conducted an interesting study. They demonstrated no enhancement of insulin secretion under low glucose concentrations and a significant increase of insulin secretion under high glucose concentrations. Their results have shown that the intravenous injection of 200 μg of the PEI/pSIGLP1N κB complex with an N/P ratio 5 could decrease blood glucose after the second injection in the second day, after which the blood glucose levels did not return to the preadministration baseline until the 17th day after injection [19]. According to the obtained results, the use of PEI as a carrier could efficiently protect and transfect the DNA plasmid to the targeted site. The most important property of PEI is the highly cationic charge density. The nitrogen atoms of PEI can be protonated in the endosomal pH and therefore, act as a proton sponge. This effect is very crucial for gene delivery to the cells and occurs by passive chloride influx into the endosomes, which leads to osmotic swelling and disruption of the cellular endosomes. The disruption of the endosomal membrane permits to escape the endocytosed PEI–DNA complexes. However, PEI is a very cytotoxic polymer which diminishes its application in clinic. Some factors which effect its cytotoxicity include molecular weight, incubation time, concentration, and the density of cationic groups [20 - 22]. Some studies try to decrease its toxicity by using different derivatives of PEI which could not completely solve this problem.

Type 1A diabetes is related to TH1/TH2 imbalance and enhancement of TH1 cytokines like IFN-γ, IL-2, and TNF-α. If TH2 cytokines increase like IL-4 and IL-10, the balance could berestored and lead to the prevention of autoimmune diabetes. Cytokine delivery has high cost and cytokine has a very short plasma half-life. In this regard, gene delivery could be very useful for cytokine expression into the human body [23, 24]. Manki and Singh in one study developed polymeric nanomicelles with a positive charge to deliver DNA plasmid encoding IL-4 and IL-10 for the prevention of autoimmune diabetes in mice [25]. They used N-acyl substitute of low molecular weight chitosan for the formulation of nanomicelles with a size range of 90 nm. In their study, oleic and linoleic acid substitutes were used to make stable complexes with DNA and protect it from enzymatic degradation and make high transfection efficiency. The polyplex formulations were injected intramuscularly as a single dose into the anterior tibialis muscle of mice then the IL-4 and IL-10 levels were measured for 6 weeks. In that study, chitosan polymer wasused with a molecular weight of 50 kDa as a gene carrier with a high positive charge density. Their results have shown polymeric nanomicelles were as effective as positive control which was related to the buffering ability and the small size of nanoparticles which causes efficient endosomal escape. N-linoleyl LMWC nanomicelles showed great efficiency compared with other treatment groups, and the expression level of IL-4 and IL-10 was significantly elevated during 6 weeks.

N- linoleyl LMWC delivery system also showed significantly low blood glucose levels compared to other groups. In their study, the biocompatibility of the delivery system was also investigated and no chronic inflammation of the injection site muscle was observed.

RNA DELIVERY APPROACH

The level of proteins related to diseases can be changed by upregulation or downregulation. By using RNA interference technology, we can selectively downregulate proteins. Short interfering RNAs (siRNA) are synthetic RNAs which can selectively degrade complementary mRNA-related to a specific protein and eventually downregulate the specific protein [26].

RNA delivery can be used for the treatment of diabetes disease. Down regulation of glutamic acid decarboxylase (GAD) expression in transgenic mice could completely protect islet β-cells and prevent the destruction of pancreatic β-cells. Researchers used a polymeric nanoparticle based on poly (ethylene glycol)-grafted poly-L-lysine (PEG-g-PLL) as a gene carrier for the delivery of antisense GAD mRNA expression plasmid (pRIP-AS-GAD). In their study, a weight ratio 1:3 was observed that could make the highest transfection efficiency [27]. Studies have shown the overexpression of cAMP response element binding protein (CREB)-regulated transcriptional coactivator 2 (CRTC2), which plays a crucial role in causing high hepatic gluconeogenesis in diabetic patients. In a study of our team, RNA interference technology was used for downregulation of CRTC2 gene expression. We used chitosan nanoparticles as an siRNA carrier which targeted glycyrrhetinic acid (GA) as a liver-specific ligand. The prepared NPs were intravenously injected to the diabetic rats by a single dose of NPs (20 μg siRNA). In the study, the plasma glucose concentration was evaluated for 7 days following the first injection. Our results have shown that the blood glucose level significantly decreased 1 day after injection, and this effect was observed until 5 days after injection [28].

In other studies, the downregulation of arachidonate 15-lipoxygenase (Alox15) which is expressed by pancreatic immune cells resulted in the prevention of T1D. In their study, both lipid-and polyethylenimine-based vehicles were used. Different routes of administration were also studied to find the most efficient route of siRNA delivery to pancreas-associated immune cells. Their results have shown that the intra-peritoneal (i.p) route is more effective compared with intravenous administration for siRNA delivery to pancreas-associated immune cells. Furthermore, it demonstrated that a single injection of siRNA/nanocarrier can downregulate the target protein for at least 7 days [29].

FUTURE AND CHALLENGES

Although nucleotide delivery by using DNA plasmid or RNA opened a new window for an efficient therapy in diabetes disease, challenges remain in the wide use of this approach to diabetes treatment. Robust analytical methods which can precisely and accurately characterize the properties of gene vehicles *in vitro* and *in vivo* are essential for the development of gene delivery systems and their scale-up [30]. Furthermore, the current investigated techniques require more characterization for use in clinic with respect to different aspects, particularly toxicity issues. In this regard, several challenges must be addressed to translate these concepts into clinical reality [31].

1. Nanoparticle Design and Biocompatibility: Designing nanoparticles that are biocompatible, stable, and can efficiently deliver genetic material to target cells is a complex task. Ensuring these nanoparticles do not elicit an immune response or cause toxicity in the body is crucial [32].

2. Targeted Delivery: Achieving targeted delivery to the specific cells and tissues affected by diabetes is vital for the success of gene therapy. Nanoparticles must be engineered to recognize and bind to these target cells, preventing off-target effects [33].

3. Gene Regulation: Fine-tuning gene expression in response to dynamic changes in the body's glucose levels is essential for maintaining optimal blood sugar levels. Controlling the amount and duration of gene expression presents a significant challenge [34].

4. Immunogenicity: The immune system's response to gene therapy must be carefully monitored and controlled. Immunogenic reactions can neutralize the therapeutic effects of gene therapy or lead to adverse effects [35].

5. Long-term Efficacy: Ensuring the long-term stability and efficacy of gene therapy in diabetes treatment is crucial. The persistence of therapeutic effects over time is necessary to provide sustainable benefits to patients [36].

CONCLUSION

In recent years, there have been many efforts made to develop delivery systems for treatment of different diseases. In this context, some mRNA-based vaccines (*e.g.*, BioNTech/Pfizer and Moderna mRNA) used nanotechnology (lipid-based nanoparticles) for the delivery of nucleotides to the host cells. Their obtained results have shown the great potential of nanotechnology for gene delivery from laboratory to clinical and industrial applications [37, 38]. The use of nano vehicles for gene delivery in diabetes is still at the lab stage. Based on their potential to control blood glucose, it seems they could be available on the market in the next decade.

REFERENCES

[1]   Rabbani M, Sadeghi HM, Moazen F, Hasanzadeh A, Imani EF, Rastegari A. Association of KCNJ11 (E23K) gene polymorphism with susceptibility to type 2 diabetes in Iranian patients. Adv Biomed Res 2015; 4(1): 1.
[http://dx.doi.org/10.4103/2277-9175.148256] [PMID: 25625107]

[2]   Rastegari A, Rabbani M, Sadeghi HM, Imani EF, Hasanzadeh A, Moazen F. Pharmacogenetic association of KCNJ11 (E23K) variant with therapeutic response to sulphonylurea (glibenclamide) in Iranian patients. Int J Diabetes Dev Ctries 2015; 35(4): 630-1.
[http://dx.doi.org/10.1007/s13410-015-0316-1]

[3]   Wolff JA, Malone RW, Williams P, *et al.* Direct gene transfer into mouse muscle *in vivo*. Science 1990; 247(4949): 1465-8.
[http://dx.doi.org/10.1126/science.1690918] [PMID: 1690918]

[4]   Sahay G, Alakhova DY, Kabanov AV. Endocytosis of nanomedicines. J Control Release 2010; 145(3): 182-95.
[http://dx.doi.org/10.1016/j.jconrel.2010.01.036] [PMID: 20226220]

[5]   Lin MK, Farrer MJ. Genetics and genomics of Parkinson's disease. Genome Med 2014; 6(6): 48.
[http://dx.doi.org/10.1186/gm566] [PMID: 25061481]

[6]   Kim SW. Polymeric gene delivery for diabetic treatment. Diabetes Metab J 2011; 35(4): 317-26.
[http://dx.doi.org/10.4093/dmj.2011.35.4.317] [PMID: 21977450]

[7]   Maestro S, Weber ND, Zabaleta N, Aldabe R, Gonzalez-Aseguinolaza G. Novel vectors and approaches for gene therapy in liver diseases. JHEP Reports 2021; 3(4): 100300.
[http://dx.doi.org/10.1016/j.jhepr.2021.100300] [PMID: 34159305]

[8]   Subramani K, Pathak S, Hosseinkhani H. Recent trends in diabetes treatment using nanotechnology. Dig J Nanomater Biostruct 2012; 7(1).

[9]   Pack DW, Hoffman AS, Pun S, Stayton PS. Design and development of polymers for gene delivery. Nat Rev Drug Discov 2005; 4(7): 581-93.
[http://dx.doi.org/10.1038/nrd1775] [PMID: 16052241]

[10]   Guo J, Cheng WP, Gu J, *et al.* Systemic delivery of therapeutic small interfering RNA using a pH-triggered amphiphilic poly-l-lysine nanocarrier to suppress prostate cancer growth in mice. Eur J Pharm Sci 2012; 45(5): 521-32.

[http://dx.doi.org/10.1016/j.ejps.2011.11.024] [PMID: 22186295]

[11]   Howard KA, Rahbek UL, Liu X, *et al.* RNA interference *in vitro* and *in vivo* using a novel chitosan/siRNA nanoparticle system. Mol Ther 2006; 14(4): 476-84.
       [http://dx.doi.org/10.1016/j.ymthe.2006.04.010] [PMID: 16829204]

[12]   Koh JJ, Ko KS, Lee M, Han S, Park JS, Kim SW. Degradable polymeric carrier for the delivery of IL-10 plasmid DNA to prevent autoimmune insulitis of NOD mice. Gene Ther 2000; 7(24): 2099-104.
       [http://dx.doi.org/10.1038/sj.gt.3301334] [PMID: 11223991]

[13]   Basarkar A, Singh J. Poly (lactide-co-glycolide)-polymethacrylate nanoparticles for intramuscular delivery of plasmid encoding interleukin-10 to prevent autoimmune diabetes in mice. Pharm Res 2009; 26(1): 72-81.
       [http://dx.doi.org/10.1007/s11095-008-9710-4] [PMID: 18779928]

[14]   Jeon HJ, Oh TK, Kim OH, Kim ST. Delivery of factor VIII gene into skeletal muscle cells using lentiviral vector. Yonsei Med J 2010; 51(1): 52-7.
       [http://dx.doi.org/10.3349/ymj.2010.51.1.52] [PMID: 20046514]

[15]   Kormann MSD, Hasenpusch G, Aneja MK, *et al.* Expression of therapeutic proteins after delivery of chemically modified mrna in mice. Nat Biotechnol 2011; 29(2): 154-7.
       [http://dx.doi.org/10.1038/nbt.1733] [PMID: 21217696]

[16]   Kessler PD, Podsakoff GM, Chen X, *et al.* Gene delivery to skeletal muscle results in sustained expression and systemic delivery of a therapeutic protein. Proc Natl Acad Sci USA 1996; 93(24): 14082-7.
       [http://dx.doi.org/10.1073/pnas.93.24.14082] [PMID: 8943064]

[17]   Drucker DJ. Minireview: The glucagon-like peptides. Endocrinology 2001; 142(2): 521-7.
       [http://dx.doi.org/10.1210/endo.142.2.7983] [PMID: 11159819]

[18]   Oh S, Lee M, Ko KS, Choi S, Kim SW. GLP-1 gene delivery for the treatment of type 2 diabetes. Mol Ther 2003; 7(4): 478-83.
       [http://dx.doi.org/10.1016/S1525-0016(03)00036-4] [PMID: 12727110]

[19]   Choi S, Oh S, Lee M, Kim SW. Glucagon-like peptide-1 plasmid construction and delivery for the treatment of type 2 diabetes. Mol Ther 2005; 12(5): 885-91.
       [http://dx.doi.org/10.1016/j.ymthe.2005.03.039] [PMID: 16039908]

[20]   Ferrari S, Moro E, Pettenazzo A, Behr JP, Zacchello F, Scarpa M. ExGen 500 is an efficient vector for gene delivery to lung epithelial cells *in vitro* and *in vivo* Gene Ther 1997; 4(10): 1100-6.
       [http://dx.doi.org/10.1038/sj.gt.3300503] [PMID: 9415317]

[21]   Fischer D, Bieber T, Li Y, Elsässer HP, Kissel T. A novel non-viral vector for dna delivery based on low molecular weight, branched polyethylenimine: Effect of molecular weight on transfection efficiency and cytotoxicity. Pharm Res 1999; 16(8): 1273-9.
       [http://dx.doi.org/10.1023/A:1014861900478] [PMID: 10468031]

[22]   Jeong JH, Song SH, Lim DW, Lee H, Park TG. DNA transfection using linear poly(ethylenimine) prepared by controlled acid hydrolysis of poly(2-ethyl-2-oxazoline). J Control Release 2001; 73(2-3): 391-9.
       [http://dx.doi.org/10.1016/S0168-3659(01)00310-8] [PMID: 11516514]

[23]   Chernajovsky Y, Gould DJ, Podhajcer OL. Gene therapy for autoimmune diseases: Quo vadis? Nat Rev Immunol 2004; 4(10): 800-11.
       [http://dx.doi.org/10.1038/nri1459] [PMID: 15459671]

[24]   Li L, Yi Z, Tisch R, Wang B. Immunotherapy of type 1 diabetes. Arch Immunol Ther Exp 2008; 56(4): 227-36.
       [http://dx.doi.org/10.1007/s00005-008-0025-2] [PMID: 18726144]

[25]   Mandke R, Singh J. Cationic nanomicelles for delivery of plasmids encoding interleukin-4 and interleukin-10 for prevention of autoimmune diabetes in mice. Pharm Res 2012; 29(3): 883-97.

[http://dx.doi.org/10.1007/s11095-011-0616-1] [PMID: 22076555]

[26] McManus MT, Sharp PA. Gene silencing in mammals by small interfering rnas. Nat Rev Genet 2002; 3(10): 737-47.
[http://dx.doi.org/10.1038/nrg908] [PMID: 12360232]

[27] Lee M, Han S, Ko KS, *et al.* Repression of gad autoantigen expression in pancreas β-cells by delivery of antisense plasmid/peg-g-pll complex. Mol Ther 2001; 4(4): 339-46.
[http://dx.doi.org/10.1006/mthe.2001.0458] [PMID: 11592837]

[28] Rastegari A, Mottaghitalab F, Dinarvand R, *et al.* Inhibiting hepatic gluconeogenesis by chitosan lactate nanoparticles containing crtc2 sirna targeted by poly(ethylene glycol)-glycyrrhetinic acid. Drug Deliv Transl Res 2019; 9(3): 694-706.
[http://dx.doi.org/10.1007/s13346-019-00618-1] [PMID: 30825078]

[29] Leconet W, Petit P, Peraldi-Roux S, Bresson D. Nonviral delivery of small interfering RNA into pancreas-associated immune cells prevents autoimmune diabetes. Mol Ther 2012; 20(12): 2315-25.
[http://dx.doi.org/10.1038/mt.2012.190] [PMID: 22990670]

[30] Wong SY, Pelet JM, Putnam D. Polymer systems for gene delivery—past, present, and future. Prog Polym Sci 2007; 32(8-9): 799-837.
[http://dx.doi.org/10.1016/j.progpolymsci.2007.05.007]

[31] Veiseh O, Tang BC, Whitehead KA, Anderson DG, Langer R. Managing diabetes with nanomedicine: Challenges and opportunities. Nat Rev Drug Discov 2015; 14(1): 45-57.
[http://dx.doi.org/10.1038/nrd4477] [PMID: 25430866]

[32] Skotland T, Iversen TG, Sandvig K. Development of nanoparticles for clinical use. Nanomedicine 2014; 9(9): 1295-9.
[http://dx.doi.org/10.2217/nnm.14.81] [PMID: 25204821]

[33] Pavani G, Amendola M. Targeted gene delivery: where to land. Front genome ed 2021.
[http://dx.doi.org/10.3389/fgeed.2020.609650]

[34] Ji J, Tao Y, Zhang X, *et al.* Dynamic changes of blood glucose, serum biochemical parameters and gene expression in response to exogenous insulin in arbor acres broilers and silky fowls. Sci Rep 2020; 10(1): 6697.
[http://dx.doi.org/10.1038/s41598-020-63549-9] [PMID: 32317707]

[35] Sack BK, Herzog RW. Evading the immune response upon *in vivo* gene therapy with viral vectors. Curr Opin Mol Ther 2009; 11(5): 493-503.
[PMID: 19806497]

[36] Flotte TR. Gene therapy: The first two decades and the current state-of-the-art. J Cell Physiol 2007; 213(2): 301-5.
[http://dx.doi.org/10.1002/jcp.21173] [PMID: 17577203]

[37] Pushparajah D, Jimenez S, Wong S, Alattas H, Nafissi N, Slavcev RA. Advances in gene-based vaccine platforms to address the COVID-19 pandemic. Adv Drug Deliv Rev 2021; 170: 113-41.
[http://dx.doi.org/10.1016/j.addr.2021.01.003] [PMID: 33422546]

[38] Park KS, Sun X, Aikins ME, Moon JJ. Non-viral COVID-19 vaccine delivery systems. Adv Drug Deliv Rev 2021; 169: 137-51.
[http://dx.doi.org/10.1016/j.addr.2020.12.008] [PMID: 33340620]

SUBJECT INDEX

A

Absorption 11, 38, 40, 58, 61, 63, 64, 65, 76, 83, 88, 89
 carbohydrate 40
 insulin-mediated glucose 11
 intestinal glucose 38
 of insulin 63, 64
 pulmonary 65
Acid(s) 6, 13, 62, 63, 64, 65, 81, 82, 84, 86, 87, 89, 102, 103, 104, 106
 bile 13
 boronic 87
 cholic 81
 folic 62
 gluconic 89
 itaconic 64
 nucleic 64, 65, 102, 103, 104
 pachymic 86
 phenylboronic 64, 65
 Polylactic 82
 polysaccharide hyaluronic 84
 targeted glycyrrhetinic 106
 tricarboxylic 6
Activity 12, 16, 37, 38, 61, 63, 64, 74, 75, 87, 88, 89, 90, 93
 anti-hypertriglyceridemia 38
 antibacterial 87
 antihyperglycemic 88
 antioxidant 74
 enzymatic 61, 63, 64, 88
 hypoglycaemic 74, 75
 insulinotropic 89
 tyrosine kinase 38
Acute 39, 44
 pancreatitis 44
 renal failure 39
Agents 42, 43, 46, 76
 emulsifying 76
 non-insulin 46
Alginate-based nanoparticles 80

Alpha-glucosidase inhibitors (AGIs) 40, 41
Amino acids 7, 13, 41
 glucogenic 7
AMP-dependent protein kinase 38
Angiogenesis 90
Anti-diabetic activity 92
Anti-oxidant functions 92
Antidiabetic therapy 93
Apoptosis 4, 10, 90, 91
 arrest 90
Atherosclerosis 9
Atherosclerotic 9, 35
 lesions 9
Autoantibodies 5, 17
 sensitive 17
Autoimmune diabetes 17, 104, 105
Autoimmunity 3, 5, 15

B

Basal insulin 32, 33, 34, 42, 46
 preparations 33
Bio-materialistic applications 81
Bioavailability of insulin 57, 58
Blood glucose 14, 57, 76, 89, 104, 105
 downregulated 76

C

Cancer, bladder 40
Carbohydrate intake 34
Carbon nanotubes (CNTs) 86, 87
Cardiac 45, 77
 dysfunction 45
 index 77
Cardiomyocytes 9
Cardiotoxicity 39
Cardiovascular 1, 16, 31, 35, 38, 39, 45
 complications 31, 39
 disease 35, 38
 disorders 1
 events, adverse 45

www.ingramcontent.com/pod-product-compliance
Lightning Source LLC
Chambersburg PA
CBHW041717210326
41598CB00007B/685